Practical Building Forms and Agreements

Quick Reference to Form

Practical Building Forms and Agreements

Andrew Pike

Solicitor of the Supreme Court and Notary Public Partner,
Simmons & Simmons, London.

E & FN SPON

An Imprint of Chapman & Hall

London · Glasgow · New York · Tokyo · Melbourne · Madras

Published by
E & FN Spon, an imprint of Chapman & Hall, 2–6 Boundary Row,
London SE1 8HN

Chapman & Hall, 2–6 Boundary Row, London SE1 8HN, UK

Blackie Academic & Professional, Wester Cleddens Road,
Bishopbriggs, Glasgow G64 2NZ, UK

Chapman & Hall Inc., 29 West 35th Street, New York NY10001, USA

Chapman & Hall Japan, Thomson Publishing Japan, Hirakawacho
Nemoto Building, 6F, 1–7–11 Hirakawa-cho, Chiyoda-ku, Tokyo 102,
Japan

Chapman & Hall Australia, Thomas Nelson Australia, 102 Dodds
Street, South Melbourne, Victoria 3205, Australia

Chapman & Hall India, R. Seshadri, 32 Second Main Road, CIT East,
Madras 600 035, India

First edition 1993

© 1993 Simmons & Simmons

Printed in Great Britain by The Alden Press Ltd, Oxford

ISBN 0 419 18150 4

A catalogue record for this book is available from the British Library

Library of Congress Cataloging-in-Publication data available

*The cover illustration shows concrete cores being slipformed at a 20
storey office development in London (courtesy of Costain Engineering
and Construction Ltd).*

∞ Printed on permanent acid-free text paper, manufactured in
accordance with the proposed ANSI/NISO Z 39.48-199X and ANSI Z
39.48-1984

To Anne, Arabella and Alexandra

SIMMONS & SIMMONS

We are one of the largest law firms in the City of London, with well over 100 partners and a total strength approaching 1,000. Since the firm's foundation in 1896, our growth has been continuous, with accelerating expansion over recent years.

We have a broadly-based commercial and financial practice with corporate work at the heart of our business. Our offices in Paris, Brussels, Lisbon, Hong Kong and New York, together with our close links with leading lawyers worldwide, reinforce the strength of our international work. We are a full-service law firm and act for clients of all descriptions: from the largest multinational companies to small businesses, from government departments to individuals.

Our aim is to provide practical advice promptly and efficiently. In order to achieve this, we emphasize direct partner involvement and offer a wide knowledge of business and commerce to supplement our legal expertise. We harness and organize our total resources to meet the requirements of our clients for quality, reliability, speed of response and value.

LONDON

14 Dominion Street
London EC2M 2RJ
Telephone 44-71-628 2020/528 9292
Facsimile 44-71-588 4129/588 9418
Telex 888562 SIMMON G

PARIS

2 Avenue Bugeaud
75116 Paris
Telephone 33-1-45016767
Facsimile 33-1-45012232
Telex TRANSAV 649381 F

BRUSSELS

Regent Building
Boulevard du Régent, 58
BTE 1
1000 Brussels
Telephone 32-2-5117270
Facsimile 32-2-5134220
Telex 22087 SIMLAW B

SIMMONS & SIMMONS

In response to growing demand, we formed a specialist Development and Construction Law Group in 1987, to provide a comprehensive range of services to employers, developers, consultants and contractors engaged in the construction and engineering industries.

The Group offers services, both domestically and internationally, in relation to project structuring and financing, tax planning, site acquisition and assembly, development agreements, building contracts and tenders, the appointment of the professional team, collateral warranties, insurance and dispute resolution.

We are founder members of the newly created Centre for Dispute Resolution (CEDR), which aims to promote, where possible, the resolution of disputes by mediation, conciliation and mini-trial, rather than more traditional arbitration and litigation. CEDR is of particular interest to those involved in construction disputes.

LISBON	**HONG KONG**	**NEW YORK**
Grupo Legal Português Rua Fialho de Almeida N°1,3.°Esq. 1000 Lisbon Telephone 351-1-3153262 Facsimile 351-1-3153263 *In association with* *J&A Garrigues, Pinheiro Neto & Co* *and F. Castelo Branco & Nobre Guedes*	24th Floor, Jardine House One Connaught Place, Central Hong Kong Telephone 852-8681131 Facsimile 852-8105040 Telex 75888 SANDS HX	115 East 57th Street New York, NY 10022 Telephone 1-212-751 9233 Facsimile 1-212-355 3594 Telex 149543

CONTENTS

Appendices

PREFACE

An Employer embarking upon a building project will inevitably have to enter into a number of complex contracts. In a conventional project, he will need to enter into terms of engagement with consultants – typically an Architect, a Quantity Surveyor, a Structural Engineer and a Building Services Engineer – and a Main Contract with the Contractor entrusted with the building Works. There are published institutional standard forms for all these purposes, but usually they will need to be amended in one way or another, as explained in the introductory chapters.

The purpose of this book is to help Employers and their advisers (especially their Quantity Surveyors) in drawing up all the contracts required on a normal UK building project, including:

- the terms of engagement of consultants;
- the building Contract with the Main Contractor, whether it be a traditional lump sum, 'design and build' or management contract;
- Trade Contracts with Trade Contractors in a construction management project (for which there are no published UK standard forms);
- the collateral warranties required by the various contractual structures;
- performance bonds; and
- personal and parent company guarantees.

The Forms included in this book are a co-ordinated attack upon the task of legal documentation for building rojects, and they are based upon, and incorporate by reference, the relevant published standard forms. Incorporation by reference keeps the Forms concise. The use of published standard forms in this way should help to reduce the time and expense consumed in negotiations and discussions, because the parties will be working from a well-known and generally accepted base and not from a one-sided document produced by one party entirely in its own interests.

The draft appointments of consultants are all based upon the forms of appointment published by the relevant professional institutions; the draft collateral warranties are all based upon the British Property Federation Form CoWa/F, Second Edition 1990; and the Main Contracts are all based upon JCT forms.

The emphasis is upon **practicality**. The Forms are intended to achieve a good and reasonable deal for the Employer, not a theoretical perfection which would be unachievable and non-negotiable in practice. There is, for example, hardly a clause in the JCT forms which could not somehow be amended in the Employer's favour, but that would produce fussy amendments of great length which might clog negotiations with Contractors, even when market conditions are adverse to Contractors. The Forms stick to basic and essential points.

In any building project, the Employer must necessarily invest a great deal of trust, not only in his consultants, but also in his Contractor – or several Contractors if it is a construction management project. The success of the project also depends very largely upon continuing good relationships between the parties. It is, therefore, hardly wise for the Employer to begin the project by attempting to oppress the other parties involved with unreasonable contractual terms. Especially in present market conditions, the Employer may well find those who will sign anything in order to obtain work. However, I suspect that such an Employer's 'victory' is likely to be temporary, and on paper only, with a fine harvest of discord, disputes and claims to follow.

Therefore, I have tried to achieve a reasonable balance, so that at least no party involved should start off with legitimate contractual grievances.

Although this book is primarily intended to help Employers, it is certainly not intended to be anti-Contractor. I was in industry as a Contractor's lawyer for 14 years until 1988 and fully understand the rigorous environment in which Contractors have to operate, especially during a depressed market. However, it is evident that the contractors' representatives on the JCT do an excellent job in protecting the interests of their constituents, and some aspects of the JCT forms give Employers legitimate cause for concern. At several points in the text where suggested amendments to the JCT forms might unduly prejudice Contractors, I have also put forward suggestions for reasonable compromise.

With regard to the professional appointments contained in this work, I have suggested that the various professional institutions' standard forms should be used with only minimal amendments.

In relation both to the JCT forms and the various institutions' forms of appointment, many of the amendments relate to matters not covered by the standard forms at all, such as the dread subjects of collateral warranties and professional indemnity insurance.

There is one piece of advice to Employers which I believe is sufficiently important to place in this Preface: if an Employer permits consultants or Contractors to begin work on an informal basis (such as a letter of intent), his bargaining position is thereafter weakened or eliminated. **All the relevant contractual documents should be formal and legal from the beginning.** Otherwise, the Employer may expect to spend months or years, at substantial legal expense, in haggling over precise terms, particularly the terms of collateral warranties.

Andrew Pike
February 1993

ACKNOWLEDGEMENTS

I wish to express my appreciation of the help given in the preparation of this book by Davis, Langdon & Everest, leading Chartered Quantity Surveyors and authors of Spon's construction price books, who kindly reviewed the manuscript and made a number of most useful points and suggestions.

I also gratefully acknowledge the help of the British Property Federation and its allied institutions, which gave permission for the reproduction in this book of their Collateral Warranties. These appear as Appendices B, C and D.

My thanks are also especially due to my partners Philip Bretherton, Philip Vaughan, Robert Bryan and all my other colleagues at Simmons & Simmons, from whom I have learnt so much since joining the firm in 1990; to Miss Lin Chester for word-processing the manuscript; and to Miss Debbie Grayling for producing the text on our 'in-house' desktop publishing facilities.

The cover photograph, which is by courtesy of Costain Engineering and Construction Limited shows concrete cores being slipformed at a 20 storey office development for Barclays Bank in Lombard Street, London, EC3.

Andrew Pike

NOTICE

All parties must rely exclusively upon their own skill and judgment, or upon those of their advisers, when making use of the information and guidance given in this work, which should not be used without taking appropriate legal advice on the matter in question. It is in the nature of all legal precedents that they can only rarely be used in a specific case without amendment or clarification. Neither the author, nor the publishers, nor the copyright holders, assume any liability to anyone for any loss or damage caused by any error or omission in the work, whether such error or omission is the result of negligence or any other cause. Any and all such liability is disclaimed.

CHAPTER 1

INTRODUCTION AND REVIEW OF THE FORMS

INTRODUCTION

1. This chapter contains a brief introduction and goes on to describe the Forms given in this book in reasonable detail. Chapter 2 explains why certain amendments and additions to the JCT forms of contract are suggested in the Forms. Chapters 3 and 4 are brief notes respectively on collateral warranties and letters of intent. It is thought that most readers of this book will have a basic knowledge of building projects and the various contractual structures in general use. However, these basic concepts are briefly explained in the author's paper 'Construction Management and the JCT Contracts' (see Chapter 5), together with charts illustrating the relationships of the several parties involved. Readers are urged to review Chapter 5, if in any doubt about these matters.

2. In the UK, we are most fortunate to have the benefit of a large range of published institutional standard forms, upon which building contracts and allied documents, such as consultants' appointments and collateral warranties, may be based. It is, therefore, unnecessary to start from a blank sheet, and the use of generally accepted standard forms serves the very useful purpose of minimizing the time spent in negotiations.

3. For appointments of consultants, such as Architects, Quantity Surveyors and Engineers, there are widely used forms published respectively by the Royal Institute of British Architects ('RIBA'), the Royal Institution of Chartered Surveyors ('RICS') and the Association of Consulting Engineers ('ACE'). However, a well-advised Employer or funding institution will often require these forms to be added to and amended. For example:

    – there is generally no provision in the published forms for collateral warranties in favour of third parties, such as funding institutions, purchasers or tenants;
    – there is generally no provision for professional indemnity insurance to be taken out and maintained by the relevant consultant; and
    – under the RIBA standard forms of Architect's appointment (both the 1982 Edition and SFA/92), the Architect may resign at any time upon reasonable notice, which may cause the Client severe difficulties and loss, especially if

the Architect has at that point received a large percentage of his fee.

Nevertheless, the great majority of the provisions of the relevant forms are reasonable and acceptable, and the forms of consultants' appointments given in this book are relatively brief documents incorporating the relevant published form by reference.

4. Building contracts in the UK are usually entered into in one of the forms published by the Joint Contracts Tribunal for the Standard Form of Building Contract ('JCT'). The constituent bodies of the JCT are listed in the glossary to Chapter 5. Again, a well-advised Employer or funding institution will very often require the JCT forms to be amended and added to in certain important respects. For example:

   – there is no provision in any of the JCT forms for collateral warranties in favour of third parties;
   – there is no provision for performance bonds or parent company guarantees, which are very frequently required by the Employer in practice;
   – there are serious pitfalls from the Employer's point of view, which are explained in Chapter 2; and
   – there is no provision in the JCT Contractor's Design Form ('JCT 81') for professional indemnity insurance to be taken out and maintained by the Contractor in respect of his design responsibilities, which will be at least equivalent to those of an Architect.

5. The preparation of building contract documentation in the UK has traditionally devolved upon Quantity Surveyors. However, the increased complexity of such documents in recent years, and particularly the proliferation of collateral warranties, has caused more and more lawyers to become involved in the construction field.

6. The great majority of the provisions of the JCT forms are reasonable and acceptable, if somewhat incomprehensible, and the forms of building contract given in this book are relatively brief documents incorporating the relevant JCT form by reference, with certain specific amendments of limited scope and length.

7. With regard to collateral warranties, the only published forms known to the author are the British Property Federation Collateral Warranty for Funding Institutions CoWa/F (in its three editions), the BPF Collateral Warranty for Purchasers and Tenants CoWa/P&T, 1992, and the Royal Incorporation of Architects in Scotland Duty of Care Agreement, 1988. The BPF forms

CoWa/F, Second Edition 1990 and Third Edition 1992, and CoWa/P&T 1992, are reproduced in Appendices B, C and D. Form CoWa/F, Third Edition 1992, was published while this book was in the course of preparation. As it contains, like SFA/92, a number of points adverse to the warrantee's interests, the author has continued to use the Second Edition, which is largely acceptable, as a drafting base for all the forms of collateral warranty given in this book. Proposed departures from the Second Edition are explained in the commentary upon Form 6. Form CoWa/F, Third Edition, and Form CoWa/P&T are much less favourable to the warrantee. The published notes to Form CoWa/F, Third Edition, and Form CoWa/P&T, themselves explain the limitations on the warrantor's liability under those forms. It is understood that a second edition of Form CoWa/P&T is imminent.

8. Therefore, most of the Forms given in this book are based upon, or incorporate by reference, a widely used and accepted published standard form, with limited amendments and additions. Incorporation by reference is, of course, a common legal technique, which has the effect of bringing into the relevant Form the entire text of the incorporated document, subject to the amendments and additions stated in the Form. **There is no need at all in such a case actually to sign or amend the published standard form itself.** This method of proceeding should save a considerable amount of bulk and expense. In view of the need to add clauses and forms of collateral warranty to the JCT forms, it is no longer very practical to follow the time-honoured method of going through the JCT printed forms, amending in manuscript and gluing in extra pages. Modern word-processing makes this practice unnecessary.

9. The Forms (which are set in a different font from these chapters) are designed for use by well-informed lawyers and non-lawyers on UK building projects. If there is a non-UK element (for example, if a party to the arrangements is incorporated or resident outside the UK) or some other peculiarity (such as an element of 'design and management'), specific legal advice on that aspect is required. However, even in such cases it is hoped that the cost and extent of that advice will be limited by the use of the Forms as a drafting base.

10. As a word of warning about using forms and agreements from a book, the author notes that, in over 22 years practising as a solicitor, he has hardly ever been able to use such a form unamended in practice. There always seems to be some unusual feature which requires special amendments. However, the author has done his best to provide all the forms and agreements required in a normal case and to make them 'user friendly'.

11. Before going on to a review of the Forms, it is necessary to mention the vital

subject of design responsibility. Employers should be aware that, in a conventional arrangement where the Employer's Architect or other consultant designs and the Contractor builds to that design, the consultant is under a professional duty of reasonable skill and care to the Employer in relation to the design, but neither the consultant nor the Contractor guarantees or warrants a favourable result, so that the building will indeed be fit for its intended purpose. For example, if the consultant specifies a building material only subsequently found to be deleterious, he will not be liable to the Employer, because he has not been negligent. The Contractor's responsibility is to follow the consultant's design and he is therefore not normally responsible for design defects. Where the Contractor assumes design responsibility under the JCT 81 'design and build' form, his design responsibility is no more than that of a consultant under a conventional arrangement, and he does **not** warrant that the Works will be fit for their intended purposes, as made known to the Contractor by the Employer's Requirements. See clause 2.5.1 of JCT 81. In view of this, some Employers may wish to request express warranties of fitness for purpose from consultants and Contractors, but strong resistance may be expected. A warranty of fitness for purpose of the relevant building will normally go beyond professional indemnity insurance coverage and is therefore extremely dangerous for the warrantor.

REVIEW OF THE FORMS

12. **Form 1** is an Architect's Appointment, incorporating by reference the RIBA Architect's Appointment, 1982 Edition. Special Conditions have been added to the RIBA Architect's Appointment relating to the following principal subjects:

    - **Professional indemnity insurance:** the Architect is required to maintain professional indemnity insurance at a stated level for a stated time. See the commentary upon clause 9 of Form 6 for further explanation. All the professional indemnity insurance clauses given in this book are based upon clause 9 of BPF Form CoWa/F, Second Edition (Appendix B).
    - **Collateral warranties:** the Architect is required to give collateral warranties to funding institutions, purchasers, tenants and others acquiring interests in the Works. Collateral warranties are further explained in Chapter 3 and in the commentary upon Form 6.
    - **Personal and/or Parent Company Guarantees:** if the Architect is an individual or a partnership, the individual or all of the partners will be personally liable for the due performance and observance of the Appointment and of the relevant collateral warranties. However, if the Architect is incorporated, i.e. a company, the Client may well require due performance and observance of the Appointment and of the collateral warranties to be

personally guaranteed. The guarantors would normally be shareholders in, and/or directors of, the incorporated Architect. If the Architect is incorporated **and** is a subsidiary of a parent company, the Client may well require the parent company to give guarantees. Of course, at least in theory, one could have both personal and parent company guarantees in respect of the same matter.

— **Duty of care to the Client:** the Architect's duty to the Client is that the Architect '**has** exercised and **will** exercise reasonable skill, care and diligence in conformity with the normal standards of the Architect's profession'. It is important that the duty should be retrospective, because the Architect will often have started work before the formal Appointment is entered into.

— **Proscribed materials:** in addition to his warranty of professional skill, care and diligence, the Architect gives a warranty about certain proscribed materials not being specified for use, or used, in the Works. The list of proscribed materials given in this and the other Forms are all as in the BPF Forms (Appendix B, C and D). Further materials may be added, if the relevant party is so advised by an expert on building materials. A lawyer should not, of course, assume responsibility for deciding which materials should be listed, as that decision will be outside his professional expertise. See also the commentary upon clause 2 of Form 6.

— **Other consultants:** it frequently happens that the Architect is the lead consultant and recommends other consultants to his Client. Sometimes, the Architect may actually employ other consultants as sub-consultants. However, that course of action will increase the Architect's legal responsibility to the Client, because the Architect will then be held responsible for the sub-consultants' work, just as any 'main contractor' will be held responsible for his 'sub-contractors'. The Architect's Appointment deals with the Architect's involvement in the appointment of other consultants by making it clear that the Architect is not responsible for other consultants employed directly by the Client, even if recommended by the Architect, but that the Architect is responsible for his sub-consultants. It is provided that the Client shall have the right of approval of sub-consultants' terms of engagement, and that sub-consultants shall give collateral warranties to the Client, so that the Client will have direct recourse to the sub-consultants, as well as the Architect, in the event of sub-consultants' negligence. However, for the reasons stated above, the Architect will usually be ill advised to employ sub-consultants.

— **Copyright:** the Client's rights regarding copyright are strengthened, by giving a wider copyright licence to the Client than that given by the RIBA Architect's Appointment, including a licence to use the relevant design documents for the **extension** of the Works. The Client is also entitled to receive copies of the relevant design documents.

— **Termination:** the Architect's right under the RIBA Architect's Appointment to resign at his discretion is removed. As mentioned above, this could be very adverse to the Client, especially if the Client had already paid a large percentage of the Architect's fee. The Architect will nevertheless be entitled, under the general law of contract, to terminate in the event of the Client's repudiatory breach of the Appointment.

13. **Form 1A** is an alternative form of Architect's Appointment, incorporating by reference the RIBA Standard Form of Agreement for the Appointment of an Architect (SFA/92), which was published during the preparation of this book. This form is much more adverse to the Client's interests than the RIBA Architect's Appointment, 1982 Edition, and the suggested amendments to SFA/92 are therefore more extensive. While the RIBA Architect's Appointment, 1982 Edition, is still accessible and has not become out-of-date because of legal or other developments, Clients have little reason to use SFA/92 in preference to the 1982 Edition.

14. **Form 2** is a Quantity Surveyor's Appointment, incorporating by reference the RICS Quantity Surveyor's Appointment.

15. **Form 3** is a Structural Engineer's Appointment, incorporating by reference ACE Agreement 3 (1984). That ACE form is intended for structural engineering work where an Architect is appointed by the Client and the form is harmonized with the RIBA Architect's Appointment incorporated by reference in Form 1.

16. **Form 4** is a Building Services Engineer's Appointment, incorporating by reference ACE Agreement 4A. That ACE form is intended for engineering services in relation to sub-contract Works, such as heating, ventilating and air conditioning services.

17. There are, of course, other published ACE forms, but ACE Agreements 3 (1984) and 4A are perhaps the two most commonly used. If it is necessary in a particular project to use one of the other ACE forms, Forms 3 and 4 would be readily adaptable for that purpose.

18. **Form 5** is a Project Manager's Appointment, incorporating by reference the RICS Project Management Agreement and Conditions of Engagement. This form is intended for use where a Project Manager (who may, of course, also be an Architect, Quantity Surveyor or Engineer) is appointed to lead the Employer's professional team. Form 5 would also be suitable for the employment of a Construction Manager in a construction management project.

19. In **Forms 1A, 2, 3, 4 and 5,** the suggested amendments to the published

6

standard forms are very much to the same substantive effect as those made in Form 1.

20. Professional appointments, such as Forms 1–5, are contracts for personal services and are therefore not assignable by either party without the other's consent. It might be considered in some cases that the Client should have the right to assign, for example, to a purchaser or mortgagee of the site, but what sensible purchaser or mortgagee would wish to employ a consultant who was not willing to be employed by him? The same point might be made in relation to 'step-in' rights under collateral warranties, but it is customary to include such rights in collateral warranties, for whatever they are worth.

21. **Form 6** is a collateral warranty to be given by a consultant to a funding institution. Like all the collateral warranties given in this book, it is based upon BPF Form CoWa/F, Second Edition 1990 (Appendix B). There is a quite extensive commentary upon Form 6, and see also Chapter 3.

22. **Form 7** is a collateral warranty to be given by a consultant to a purchaser, tenant or other 'Acquirer'. It is similar to Form 6, except that the Acquirer is naturally not given rights to 'step-in' as the consultant's substitute Client.

23. **Forms 8 and 9** are personal guarantees and a parent company guarantee to be given in respect of collateral warranties entered into by an incorporated consultant. They are very similar to Forms 10 and 11, next described.

24. **Form 10** is a personal guarantee to be given by individuals in respect of the appointment of an incorporated consultant. The form is suitable for use where there are two or more guarantors. If there is only one, the form should be converted from plural to singular and the words 'jointly and severally' should be omitted. As in all guarantees, it is most important to provide that alterations in the consultant's appointment, and other departures therefrom, shall not release the guarantors. Otherwise, they would be released under the general law. It should be noted that all forms of guarantee given in this book are in the most basic and simple terms. **They will not be suitable in the case of a complex project, or where there are foreign guarantors.**

25. **Form 11** is a parent company guarantee to be given in respect of an incorporated consultant which is also a subsidiary of a parent company.

26. **Forms 12, 13 and 14** are Main Contracts incorporating by reference respectively JCT 80 With Quantities, With Approximate Quantities and Without Quantities and, if appropriate, the JCT Sectional Completion Supplement. The suggested

amendments to this and the other JCT forms of contract are explained in Chapter 2 and in the commentary upon Form 12. It is to be noted that **Nominated Sub-Contractors**, mentioned in that commentary, only appear in JCT 80. The author has received the impression that nomination of sub-contractors is falling into a degree of disuse, at least with experienced Employers, because of the perceived disadvantages of nomination to the Employer. Nevertheless, JCT Amendment 10 issued in March 1991 made complex changes of a procedural nature relating to nomination of sub-contractors. Sometimes, Contracts embody strange creatures, such as Domestic Sub-Contractors appointed by or on behalf of the Employer, with the intention that the Employer shall, in fact, be able to choose the relevant sub-contractors, without their becoming Nominated Sub-Contractors. This may be attempted by reducing the Contractor's choice of sub-contractors under clause 19.3 from three to one, and re-hashing in some form the very lengthy clause 35 of JCT 80. The author is not in favour of such laborious and artificial methods. If the Employer wishes to nominate, he may do so, and remove the principal disadvantages of such nomination by a few simple amendments. The elaborate but necessary machinery of clause 35 may then be left in place, including the use of the ancillary NSC documents listed in clause 35.4. However, it is becoming increasingly recognized that the Employer will usually be better served by avoiding nomination altogether.

27. **Form 15** is a Main Contract incorporating by reference the JCT 81 'design and build' form. A collateral warranty to be given by consultants employed by the 'design and build' Contractor, a performance bond and a parent company guarantee are included in Form 15, because those Annexes would not be suitable for inclusion in any of the other Forms. Consultants employed by a 'design and build' Contractor are a special type of 'sub-contractor', and the relevant collateral warranty therefore bears a close resemblance to Form 21. The JCT does not publish a Sectional Completion Supplement for use with JCT 81, but the JCT Sectional Completion Supplement for use with JCT 80 is readily adaptable for use with JCT 81. As the 'design and build' Contractor assumes a professional responsibility for design pursuant to clause 2.5.1 of JCT 81, a professional indemnity insurance clause, similar to Clause 9 of Form 6, is included in Form 15. See paragraph 29 of Chapter 5 with regard to 'design and build'. The Employer and his advisers should thoroughly understand that they cannot interfere with the Contractor's design (so long as the Contractor is acting within the parameters set by the Employer's Requirements and Contractor's Proposals) without creating variations, and paying extra accordingly. If that is not acceptable to the Employer, 'design and build' should not be used.

28. **Form 16** is a Main Contract incorporating by reference the JCT Management Contract 1987 and, if required, the JCT Phased Completion Supplement for use with JCT 87. Paragraph 31 of Chapter 5 quotes JCT Practice Note MC/1, which states the circumstances in which JCT 87 is considered suitable for use. The structure of JCT 87 is quite different from the lump sum JCT 80 and JCT 81, as explained in Chapter 5, JCT 87 being basically a 'cost plus' Contract. A performance bond and parent company guarantee are included in Form 16, because those Annexes would not be suitable for inclusion in any of the other Forms. Clause 3.6 of JCT 87, relating to acceleration of completion, is a particular feature of JCT 87.

29. **Form 17** is a Main Contract incorporating the JCT Intermediate Form of Contract 1984 and, if required, the JCT IFC/SCS Sectional Completion Supplement. A performance bond is included in Form 17, because that Annexe would not be suitable for inclusion in any of the other Forms. In practice, IFC 84 is often used on quite large projects. It may be used with or without Bills of Quantities.

30. **Named Sub-Contractors** under IFC 84 have some of the disadvantages to the Employer of Nominated Sub-Contractors under JCT 80. Optional amendments are included in Form 17 in order to tackle these disadvantages, if the Employer feels that he really must 'name' sub-contractors.

31. **Form 18** is a Main Contract incorporating the JCT Agreement for Minor Works 1980. There is no JCT Sectional or Phased Completion Supplement for use with MW 80. If sectional or phased completion for a relatively small project is required, it would be better to use IFC 84, with the IFC/SCS Sectional Completion Supplement. A performance bond is included in Form 18, because that Annexe would not be suitable for inclusion in any of the other Forms. MW 80 is not suitable for use with Bills of Quantities, while IFC 84 is suitable for use with or without Bills.

32. **Forms 19 and 20** are collateral warranties to be given by Contractors under Forms 12–18 to Funds and Acquirers. These collateral warranties are similar to Forms 6 and 7 respectively, with material differences noted in the relevant commentaries.

33. **Form 21** is a collateral warranty to be given by sub-contractors under Forms 12–18 to the Employer, Fund and Acquirers.

34. **Forms 22, 23 and 24** are Construction Management Trade Contracts incorporating by reference respectively JCT 80 With Quantities, With

Approximate Quantities and Without Quantities and, if appropriate, the JCT Sectional Completion Supplement. The author is not aware of any previously published Construction Management Trade Contracts, apart from the 'CM' series of documents published by The American Institute of Architects. See Chapter 5 and the commentaries upon Forms 22, 23 and 24 for an explanation of these Forms. In a construction management project, the JCT 80 Without Quantities version will often be appropriate. Forms 22, 23 and 24 are similar to Forms 12, 13 and 14, but with a number of construction management amendments added. Form 5 might be suitable for the appointment of a Construction Manager.

35. There has been much recent discussion about the respective merits of management contracting and construction management. Chapter 5 consists of a paper which the author prepared upon that subject in 1991. The paper also refers to other source material, especially the Reading University report. Informed opinion appears to be moving to the view that, in a project which is suitable either for management contracting or construction management, construction management is better from the Employer's point of view. **However, for the vast majority of projects, neither management contracting nor construction management will be suitable.**

36. **Forms 25, 26 and 27** are collateral warranties for use in construction management projects in conjunction with Forms 22, 23 and 24. These collateral warranties are very similar in effect to Forms 19, 20 and 21.

37. **Forms 28 and 29** are parent company guarantees for the performance of collateral warranties, for use with Forms 19, 20, 21, 25, 26 and 27.

38. **Form 30** is a performance bond for use with Forms 12, 13, 14, 22, 23 and 24, i.e. the Forms incorporating JCT 80 in its various forms. Particularly in view of the frequency of insolvency in the construction industry, performance bonding in respect of the Contractor (usually 10% of the Contract Sum and often reducing to 5% on practical completion of the Works) should always be considered by the Employer. All forms of performance bond given in this book provide for their release after the expiration of the Defects Liability Period and after the Contractor has discharged all his obligations, including the payment to the Employer of any amount due to him, e.g. liquidated damages for delay. The forms of performance bond also provide for proportionate reduction upon completion of Sections or Phases, in cases where the relevant Main Contract provides for sectional or phased completion of the Works. Performance bonding in respect of consultants is most unusual. Performance bonding in respect of collateral warranties is not usual or practicable, in view of the long

life of collateral warranties, and in view of the fact that periodic fees are normally payable to the bonding institution throughout the life of the relevant bond.

39. All parties should also, when dealing with a company (whether that company is an Employer, a Contractor or a consultant), consider requiring personal guarantees from shareholders and/or directors, and a parent company guarantee if the company is a subsidiary. Unlike performance bonds, such guarantees should involve no fees and do not have to be released soon after the completion of the relevant project.

40. **Form 31** is a parent company guarantee for Contract performance, for use with Forms 12, 13, 14, 17, 18, 22, 23 and 24, i.e. the Forms incorporating JCT 80 in its various forms, IFC 84 or MW 80.

41. The final Form, **Form 32**, is a very simple form of novation agreement, intended for use where the Employer agrees to release the original Contractor from the relevant Main Contract, and to accept a substituted Contractor in the original Contractor's place. This operation may be required, for example, when the Contractor's parent company wishes to transfer the Contract to another subsidiary. In that case, as noted in the commentary upon Form 32, fresh performance bonds or parent company guarantees will be needed in respect of the substituted Contractor. Specific legal advice should be taken in respect of proposed novation.

42. Each Form is followed by its own brief commentary. Points common to several Forms are dealt with in these introductory chapters and the individual commentaries then refer back to the relevant paragraphs of these chapters. This is obviously more economical than repeating the same comment again and again after the relevant Forms.

43. The Forms given in this book are all formal agreements made by Deed, not agreements by correspondence. Now that it is normal practice for such arrangements as the engagements of consultants to include long-term obligations to maintain professional indemnity insurance and to give collateral warranties to third parties, it is most inappropriate for such arrangements to be made by correspondence.

44. The Employer or Client should require building contracts, professional appointments and collateral warranties to be executed as Deeds, in order to impose a 12 year, rather than a 6 year, limitation period. See the Limitation Act 1980, sections 5 and 8(1). In a field where latent defects may not become manifest for a long time, this is an important practical consideration, although there seems to be no logical reason why a longer limitation period should apply to Deeds.

45. Documentary formalities and methods of attestation (that is, the signature and witnessing of documents) are explained in Appendix A.

CHAPTER 2

JCT FORMS OF CONTRACT

1. Forms 12–18 and 22–24 are based upon five JCT forms, counting JCT80 as one form for this purpose. JCT 80, IFC 84 and MW 80 all represent the conventional or traditional contractual arrangement, where the Architect designs and the Contractor builds to that design. JCT 81 represents the 'fast track' or 'design and build' concept, where the Contractor takes over the design function of the Architect. Under JCT 87, the Architect designs and the Management Contractor builds to that design as in JCT 80, but the Management Contractor is remunerated by a fee, and the Contract is entered into on the basis of an estimate of prime cost rather than an ascertained Contract Sum. The Management Contractor is also relieved of a very great deal of his risk and responsibility, especially with regard to the Works Contractors who actually execute the Works as sub-contractors to the Management Contractor.

2. There are two types of building contract **not** represented in the JCT forms, namely 'construction management', where there is no single Main Contractor, the Employer enters into a number of direct contracts with Contractors for the various works packages and appoints a Construction Manager to oversee and co-ordinate the Works (see Chapter 5); and 'design and management', where the Contractor undertakes responsibilities similar to those of a Management Contractor under JCT 87, but in addition assumes design responsibility similar to that of a 'design and build' Contractor under JCT 81. Both of these types of contract require specially drafted documentation. Forms 22–24 are Construction Management Trade Contracts based upon JCT 80. The author has encountered the concept of 'design and management' in practice, but it is relatively rare, and would require 'one off' documentation.

3. The JCT forms of contract, and many other allied publications explaining and facilitating their interpretation and use, are available from RIBA Publications Limited. If in doubt as to which JCT form to use in a particular context, readers should consult the JCT Practice Note 20 'Deciding on the Appropriate Form of JCT Main Contract'. Readers should be aware that it is sometimes rather difficult correctly to identify by reference the relevant up-to-date JCT document, in view of re-printing, amendment and correction which take place from time to time. For example, the latest printing of JCT 80 Private With Quantities was (at the time of writing) dated February 1991, incorporating Amendments 1

(January 1984), 2 (November 1986), 4 (July 1987), 5 (January 1988), 6 (July 1988), 7 (July 1988), 8 (April 1989) and 9 (July 1990). (Amendment 3 (March 1987) applied to Without Quantities versions only). Amendments 10 (March 1991) and 11 (July 1992, corrected in September 1992) have since been published, and are available as separate documents. However, reference to the list of incorporated amendments at the back of the Contract, as reprinted in February 1991, indicates that corrections were also made and incorporated in the reprint dated May 1983 (i.e., before Amendment 1 was issued in January 1984); that corrections to Amendments 7 and 9 were made and incorporated in the reprint dated February 1991; and that other (minor) miscellaneous amendments and corrections were also made and incorporated in the reprint dated February 1991. Therefore, it seems that amendment and correction may occur during reprinting, as well as by numbered JCT Amendments. In that case, it appears to the author that it is wise to stipulate in the Contract just what printing is being referred to, in order to avoid any possible ambiguity, and the Forms relating to JCT documents (12–18 and 22–24) provide a blank for that purpose.

4. Each of the JCT-based Forms given in this book has its own commentary, but certain suggested amendments are common to more than one of them and these are noted in the following paragraphs. JCT 80, the basic form, is usually quoted in these paragraphs. Where other JCT forms are materially different, this is mentioned.

5. **Replacement of Architect, Contract Administrator or Quantity Surveyor (JCT 80 and IFC 84, Articles 3 and 4; JCT 87, Articles 3A, 3B and 4)** Both Articles 3 and 4 of JCT 80 provide that, if the relevant consultant dies or ceases to act for the purpose of the Contract, he shall be replaced by 'such other person as the Employer shall nominate within a reasonable time but in any case no later than 21 days after such death or cessation for that purpose'. The Employer's right of replacement is then qualified by the proviso 'not being a person to whom the Contractor no later than 7 days after such nomination shall object for reasons considered to be sufficient by an Arbitrator appointed in accordance with article 5'. It should be noted that some published forms of professional appointment enable the consultant to terminate the appointment by notice. If the relevant consultant is employed on such a basis, he may therefore resign unilaterally, and the Employer may be compelled to operate Articles 3 or 4. It sounds reasonable enough that the Contractor should have a right of objection to the substitute, but what is meant to happen in the long interval before arbitration? Any arbitration will probably not determine the issue until some time after the Works have been completed, but the Employer must, of practical necessity, have an Architect and Quantity Surveyor

continuously in office. Failure on the Employer's part to appoint substitutes would also be a breach of Articles 3 or 4 (see also *Croudace v London Borough of Lambeth* (1986) 33 BLR 20; 6 Con LR 70). Therefore, the Contractor's right to object, subject to later arbitration, could be a great impediment to the Employer, practically amounting to a Contractor's right of veto. The Employer may well not consider this to be acceptable, and require a free hand in the appointment of substitutes. As a compromise, the Employer might agree to appoint as substitutes only firms of a certain standing measured, for example, by numbers of partners or employees. This would be some reassurance to the Contractor that any substitute would be unlikely to act unreasonably.

6. **Successors and Assigns of the Employer (JCT 80, 81 and 87, clause 1.3)**
    The definition of 'Employer' is extended to include his successors and assigns. This is important in relation to the Employer's right to assign the Contract. If the Contract is to be assignable by the Employer, as to which see below, it should be made clear that the definition of Employer includes his successors and assigns, so that they also may assign, if necessary.

7. **Priority of Contract Documents (JCT 80, clause 2.2.1; JCT 81, clause 2.2; IFC 84, clause 1.3; MW80, clause 4.1)** Clause 2.2.1 of JCT 80 reads as follows:

    'Nothing contained in the Contract Bills shall override or modify the application or interpretation of that which is contained in the Articles of Agreement, the Conditions or the Appendix.'

    If the Employer and his Quantity Surveyor are **sure** that there is nothing contained in the Contract Bills which could fall within clause 2.2.1, then it may be allowed to stand. However, if there are any matters in the Contract Bills which could fall within clause 2.2.1 (and there will be in very many Bills of Quantities), then it should be deleted. There is a Court of Appeal decision (*Moody v. Ellis* (1983) 26 BLR 39) in which it was decided that the Bills could impose **additional** obligations, so long as they did not **override** or **modify** the printed documents, but that could be a very fine distinction in practice. The relevant sub-clauses will usually be contrary to the parties' true intentions. If the parties wish to establish an order of priority they may, of course, do so, but it is submitted that this should be a conscious decision, not taken by default.

8. **Employer's Assignment of the Contract (JCT 80, clause 19; JCT 81, clause 18; IFC 84 and MW 80, clause 3.1; JCT 87, clauses 3.19 and 3.20)**
    Clause 19.1.1 of JCT 80 and the first sentence of clause 19.1.2 read as follows:

'19.1.1 Neither the Employer nor the Contractor shall, without the written consent of the other, assign this Contract.

19.1.2 Where clause 19.1.2 is stated in the Appendix to apply then, in the event of transfer by the Employer of his freehold or leasehold interest in, or of a grant by the Employer of a leasehold interest in, the whole of the premises comprising the Works, the Employer may at any time after Practical Completion of the Works assign to any such transferee or lessee the right to bring proceedings in the name of the Employer (whether by arbitration or litigation) to enforce any of the terms of this contract made for the benefit of the Employer hereunder.'

Clause 19.1.1 gives each party a right of veto in respect of the assignment of the Contract by the other. Clause 19.1.2, where stated in the Appendix to apply, creates an exception in favour of the Employer, subject, however, to a number of limitations. Assignment under clause 19.1.2 may only be to a purchaser or tenant of the **whole**, can only take place **after** Practical Completion of the Works, and can only be of a right to bring proceedings in the name of the **Employer**. Therefore, in the common situation where the Employer may wish to sell during the course of construction, he cannot, without the consent of the Contractor, assign the Contract. Many Employers will find this unacceptable, and so will their respective funding institutions. It needs to be borne in mind that, if the Employer is given greater latitude to assign the Contract without the Contractor's consent, such an assignment would not have the effect of relieving the original Employer of **liability** to the Contractor. In the absence of the Contractor's specific agreement, the Employer may only assign his own 'property', namely his own rights under the Contract. He cannot assign the burden of the Contract and so compel the Contractor to look to the assignee for payment, without the Contractor's consent, which would normally be expressed in a novation agreement: see Form 32. Therefore, the Forms give the Employer a free hand to assign the Contract without the Contractor's consent. If the Employer is to be given a free hand to assign the Contract, the definition of 'Employer' in clause 1.3 should be extended to include his successors and assigns. If the definition is so extended, such successors and assigns will become the Employer for all purposes and will, for example, be able to assign the Contract further, if necessary. In practice, the Employer may be able to accept that he should only be able to assign to a limited class of assignee – for example, a funding institution – depending upon the circumstances of the case.

9. *Helstan Securities Limited v Hertfordshire County Council* [1978] 3 All ER 262 and the more recent cases of *Linden Gardens Trust Limited v Lenesta Sludge*

Disposals Limited and *St Martin's Corporation Limited v Sir Robert McAlpine Limited* (1990) 52 BLR 93; (1992) 57 BLR 57, emphasize that restrictions on assignability cannot easily be circumvented. It is submitted that the *Linden Gardens* and *St Martin's* cases demonstrate that the effect of prohibiting the Employer from assigning the Contract is unreasonable and that such a provision may lead Employers and their assignees into a morass. The cases turned upon the interpretation of clause 17 of JCT 63 (**not** JCT 80) and clause 3.1 of MW 80. Clause 17 of JCT 63 stated:

'(1)   The Employer shall not without the written consent of the Contractor assign this contract.

(2)   The Contractor shall not without the written consent of the Employer assign this contract, and shall not without the written consent of the Architect (which consent shall not be unreasonably withheld to the prejudice of the Contractor) sub-let any portion of the Works....'

The Court of Appeal held (to quote the BLR headnote) that:

'(Staughton LJ dissenting in part) Although the effect of clause 17 of the JCT Standard form and clause 3.1 of JCT Minor Works form was to prevent the assignment of the benefit of the contract without consent so that in the absence of the contractor's consent an assignee could not sue upon the contract, neither precluded the assignment of benefits arising under the contracts such as accrued causes of action for damages. Therefore since in the *Linden Gardens* case (but not *St. Martins*) breaches of contract had accrued before the assignment the rights of action for damages (being claims only for the fruits of performance) were validly assigned. *Per* Staughton LJ, dissenting,..... the words "assign this contract" in clause 17(2) do indeed refer to the delegation of the contractor's obligations by sub-contracting them all. They do not include assignment (in its legal sense) of the benefit of the contract or part of it....The words "assign this contract" in clause 17(1) must be given the same meaning. They prohibit vicarious performance by the employer without consent and do not prohibit assignment by the employer of the benefit of the contract or part of it.'

It is understood that an appeal to the House of Lords is in progress, and therefore the above may well not be the last word on the matter.

10. **Damages for Non-Completion (JCT 80 and 81, clause 24; IFC84, clause 2.7; JCT 87, clause 2.10)** Clause 24.2.1 of JCT 80 requires the Employer to require in writing, not later than the date of the Final Certificate, that liquidated

and ascertained damages for delay shall be paid or allowed by the Contractor. This requirement appears to serve no useful purpose, and could be a pitfall for the Employer. Therefore, it is suggested that the Employer should delete it. There has been a recent case, *Jarvis Brent Limited v Rowlinson Construction Limited* (1990) 6 Const LJ 292, which decided that such a requirement was not a condition precedent, but it would be much better from the Employer's point of view not to have the provision in the Contract in the first place.

11. **Extension of Time (JCT 80 and 81, clause 25.4; IFC 84, clauses 2.4.10 and 2.4.11; MW 80, clause 2.2; JCT 87, clause 2.13)** Clause 25.4.7 of JCT 80 entitles the Contractor to extension of time for completion of the Works upon the grounds of:

    'delay on the part of Nominated Sub-Contractors or Nominated Suppliers which the Contractor has taken all practicable steps to avoid or reduce'.

    This is the first of the major reliefs given to the Contractor in respect of Nominated Sub-Contractors, which is discussed in the commentary upon Form 12.

12. Clause 25.4.10 of JCT 80 entitles the Contractor to extension of time upon the grounds of:

    '25.4.10.1 the Contractor's inability for reasons beyond his control and which he could not reasonably have foreseen at the Base Date to secure such labour as is essential to the proper carrying out of the Works; or

    25.4.10.2 the Contractor's inability for reasons beyond his control and which he could not reasonably have foreseen at the Base Date to secure such goods or materials as are essential to the proper carrying out of the Works;'

    It is normally considered that these are matters which should be at the risk of the Contractor, rather than the Employer, and therefore clause 25.4.10 is customarily deleted. In the case of JCT 87, the equivalent provision is clause 2.10.10 of the Works Contract Conditions (Works Contract/2). See the suggested amendment to clause 2.13 of JCT 87 in Form 16, which would entail the deletion of clause 2.10.10 of the Works Contract Conditions pursuant to clause 8.2.1.1 of JCT 87, as amended by Form 16.

13. The Employer's own breach should be added as a ground for extension of time,

as it is already in clause 2.13 of JCT 87. In the absence of such a provision, the lack of contractual power to extend time by reason of the Employer's breach might result in time for completion, and damages for delay in completion, becoming 'at large' – that is, the contractual machinery for imposing liquidated and ascertained damages for delay might become inoperable. See, for example, *Rapid Building Group Limited v Ealing Family Housing Association Limited* (1984) 29 BLR 5, CA.

14.   Amendment of the **printed text** of clause 25 may have an impact on fluctuations, referred to below. The somewhat labyrinthine relationship between amendment of the printed text of clause 25 and fluctuations could be a serious pitfall for the Employer.

15.   **Determination (JCT 80, clauses 27, 28 and 28A; JCT 81, clause 27; IFC 84, MW80 and JCT 87, clause 7)** JCT Amendment 11, which was issued in July 1992 and corrected in September 1992, replaced clauses 27, 28 and 28A of JCT 80, but does not affect the other JCT forms. Amendment 11 removed one anomaly which previously existed in clauses 27 and 28 of JCT 80. These clauses provide *inter alia* for determination of the Contractor's employment in a number of eventualities, either amounting to insolvency, or closely related to insolvency. The clauses previously provided for determination upon an application being made for the appointment of an administrator, rather than upon the actual appointment of an administrator. Under former clause 27.2, determination of the Contractor's employment was automatic, if an application for the appointment of an administrator was made in respect of the Contractor. Under former clause 28.1.4, an application for the appointment of an administrator in respect of the Employer gave the Contractor an optional right to determine his own employment. These provisions were anomalous, as it is by no means impossible for unjustified applications for the appointment of an administrator to be made. Such an application may be made not only by the relevant company itself, or by its directors, but by any creditor, including any contingent or prospective creditor. The equivalent clauses 27.3.1 and 28.3.1, as substituted by Amendment 11, require that an administrator shall actually be appointed. Pending further amendment, the anomaly still exists in the other JCT forms.

16.   However, Amendment 11 has retained at least one problem for the Employer which existed in the former JCT 80 clauses, and has created some new ones. The retained problem relates to clause 27.2, which enables the Employer to determine the Contractor's employment in certain cases of serious breach, namely unjustified suspension of the Works, failing 'to proceed regularly and diligently with the Works', refusing to remove defective work or materials, or

unauthorized assignment or sub-contracting contrary to clause 19. In each case, the Architect must give the Contractor a 14 day warning notice, and the Employer's notice of determination must be given within 10 days of the expiry of the Architect's warning notice. However, clause 27.2.4 provides that the Employer's notice of determination 'shall not be given unreasonably or vexatiously'. This reasonable-sounding provision has the effect that the Employer can never be sure of his ground when determining the Contractor's employment. It is suggested that, once the Contractor's breach has occurred and the Architect's warning notice has been disobeyed, the Employer should have a clear right to determine the Contractor's employment, and therefore that clause 27.2.4 should be deleted. As a compromise, the similar fetter imposed by clause 28.2.5 on the Contractor's right of determination under clause 28.2 might also be deleted.

17. Amendment 11 also introduces certain new and surprising concessions to the Contractor's interests, which the Employer may consider to be unacceptable. The concessions, which could be commercially significant but are somewhat complex and labyrinthine in nature, are described in the following paragraphs.

18. Clause 27.3.1 lists a number of insolvency or insolvency-related events in relation to the Contractor, which may result in the determination of the Contractor's employment. If the Contractor:

'makes a composition or arrangement with his creditors, or becomes bankrupt, or

being a company,

makes a proposal for a voluntary arrangement for a composition of debts or scheme of arrangement to be approved in accordance with the Companies Act 1985 or the Insolvency Act 1986 as the case may be or any amendment or re-enactment thereof, or

has a provisional liquidator appointed, or

has a winding-up order made, or

passes a resolution for voluntary winding-up (except for the purposes of amalgamation or reconstruction), or

under the Insolvency Act 1986 or any amendment or re-enactment thereof has an administrator or an administrative receiver appointed' then such determination may follow. Clause 27.3.3 provides that where a provisional liquidator or trustee in bankruptcy is appointed, or a winding-up order is made, or the Contractor passes a resolution for voluntary winding-up (except for the purposes of amalgamation or reconstruction), the employment of the Contractor under the Contract shall be automatically determined. Clause 27.3.4 provides further that, **where clause 27.3.3 does not apply**, the Employer may at any time, unless an agreement to which clause 27.5.2.1 refers has been made, by notice to the Contractor determine the employment of the Contractor. Therefore, determination is **automatic** if the events listed in clause 27.3.3 occur, but **at the Employer's option** under clause 27.3.4 if any of the **other** events listed in clause 27.3.1 occur.

Clause 27.5 goes on to provide:

'27.5   Clauses 27.5.1 to 27.5.4 are **only applicable where clause 27.3.4 applies.**

27.5.1   From the date when, under clause 27.3.4, the Employer could first give notice to determine the employment of the Contractor, the Employer, subject to clause 27.5.3, shall not be bound by any provisions of this Contract to make any further payment thereunder and the Contractor shall not be bound to continue to carry out and complete the Works in compliance with clause 2.1.

27.5.2   Clause 27.5.1 shall apply until

either.1 the Employer makes an agreement (a '27.5.2.1 agreement') with the Contractor on the continuation or novation or conditional novation of this Contract, in which case this Contract shall be subject to the terms set out in the 27.5.2.1 agreement

or   .2 the Employer determines the employment of the Contractor under this Contract in accordance with clause 27.3.4, in which case the provisions of clause 27.6 or clause 27.7 shall apply.

27.5.3   Notwithstanding clause 27.5.1, in the period before either a 27.5.2.1 agreement is made or the Employer under clause 27.3.4 determines the employment of the Contractor, the Employer and the Contractor may make an interim arrangement for work to be carried out. **Subject to clause 27.5.4 any right of set-off which the Employer may have shall not be exercisable in respect of any payment due from the Employer to the Contractor under such interim arrangement.**

27.5.4 From the date when, under clause 27.3.4, the Employer may first determine the employment of the Contractor (but subject to any agreement made pursuant to clause 27.5.2.1 or arrangement made pursuant to clause 27.5.3) the Employer may take reasonable measures to ensure that Site Materials, the site and the Works are adequately protected and that Site Materials are retained in, on the site of, or adjacent to the Works as the case may be. The Contractor shall allow and shall in no way hinder or delay the taking of the aforesaid measures. The Employer may deduct the reasonable cost of taking such measures from any monies due or to become due to the Contractor under this Contract (including any amount due under an agreement to which clause 27.5.2.1, or under an interim arrangement to which clause 27.5.3, refers) or may recover the same from the Contractor as a debt.'

The last sentence of clause 27.5.3 provides for **automatic** suspension of the Employer's right of set-off in respect of payments due under an 'interim arrangement'. It is suggested that this constitutes a trap for an unwary Employer, and therefore that the sentence should be deleted. The 'interim arrangement' should, if properly drafted, expressly deal with the Employer's right of set-off.

19. Clause 27.6 provides for the results of the determination of the Contractor's employment under various provisions of clause 27, including clauses 27.3.3 and 27.3.4. Clause 27.6.4 provides as follows:

'27.6.4.1 Subject to clauses 27.5.3 and 27.6.4.2 the provisions of this Contract which require any further payment or any release or further release of Retention to the Contractor shall not apply; **provided that clause 27.6.4.1 shall not be construed so as to prevent the enforcement by the Contractor of any rights under this Contract in respect of amounts properly due to be discharged by the Employer to the Contractor which the Employer has unreasonably not discharged and which, where clause 27.3.4 applies, have accrued 28 days or more before the date when under clause 27.3.4 the Employer could first give notice to determine the employment of the Contractor or, where clause 27.3.4 does not apply, which have accrued 28 days or more before the date of determination of the employment of the Contractor.**

.4.2 Upon the completion of the Works and the making good of defects as referred to in clause 27.6.1 (but subject where relevant, to the exercise of the right under clause 17.2 and/or clause 17.3 of the Architect, with the consent of the Employer, not to require defects of the kind referred to in clause 17 to be made good) then within a reasonable time thereafter an account in respect of the matters referred to in clause 27.6.5 shall be set out in a statement either prepared by the Employer or in a certificate issued by the Architect.'

The new proviso to clause 27.6.4.1, by introducing a test of reasonableness, would, like clause 27.2.4, mean that the Employer could never be sure of his ground when withholding payment after determination of the Contractor's employment, and therefore it is suggested that it should be deleted.

20. Clause 27.7 provides as follows:

'27.7.1 If the Employer decides after the determination of the employment of the Contractor **not** to have the Works carried out and completed, **he shall so notify the Contractor in writing within 6 months from the date of such determination**. Within a reasonable time from the date of such written notification the Employer shall send to the Contractor a statement of account setting out:

.1.1 the total value of work properly executed at the date of determination of the employment of the Contractor, such value to be ascertained in accordance with the Conditions as if the employment of the Contractor had not been determined together with any amounts due to the Contractor under the Conditions not included in such total value;

.1.2 the amount of any expenses properly incurred by the Employer and of any direct loss and/or damage caused to the Employer as a result of the determination.

After taking into account amounts previously paid to or otherwise discharged in favour of the Contractor under this Contract, if the amount stated under clause 27.7.1.2 exceeds or is less than the amount stated under clause 27.7.1.1 the difference shall be a debt payable by the Contractor to the Employer or by the Employer to the Contractor as the case may be.

27.7.2 If after the expiry of the 6 month period referred to in clause 27.7.1 the Employer has **not** begun to operate the provisions of clause 27.6.1 and has **not** given a written notification pursuant to clause 27.7.1 the

Contractor may require by notice in writing to the Employer that he states whether clauses 27.6.1 to 27.6.6 are to apply and, if not to apply, require that a statement of account pursuant to clause 27.7.1 be prepared by the Employer for submission to the Contractor.'

This has the effect of imposing a time limit on the Employer's decision whether or not to 'build out' the Works. Clause 27.6.1 provides for the Employer to employ others to complete the Works. There seems little reason, in the present market, for the Employer to put up with such a time limit, and such a limit might place undue pressure on the Employer. Therefore, it is suggested either that clause 27.7 should be deleted, or that its time limits should be extended from 6 to 12 months.

21. Clause 28A.1.1 provides for the determination of the Contractor's employment by reason of prolonged suspension of the Works resulting from certain 'neutral' events - that is, events which are neither party's fault, namely:

'.1.1   force majeure; or

.1.2   loss or damage to the Works occasioned by any one or more of the Specified Perils; or

.1.3   civil commotion; or

.1.4   Architect's instructions issued under clause 2.3, 13.2 or 23.2 which have been issued as a result of the negligence or default of any local authority or statutory undertaker executing work solely in pursuance of its statutory obligations; or

.1.5   hostilities involving the United Kingdom (whether war be declared or not); or

.1.6   terrorist activity.'

Clause 2.3, 13.2 and 23.2, referred to in clause 28A.1.1.4, refer respectively to discrepancies in or divergences between contractual documents, variation in the Works, and postponement.

By virtue of clause 28A.6, where determination occurs by reason of 'loss or damage to the Works occasioned by any one or more of the Specified Perils.... caused by some negligence or default of the Employer or of any person for whom the Employer is responsible....' clause 28A.5.5 awards the Contractor:

'any direct loss and/or damage caused to the Contractor by the determination'.

Contrary to the previous clause 28A, this will include the Contractor's prospective loss of profit on the entire Contract, and it is suggested that Clause 28A.5.5 should be deleted.

22. **Retention (JCT 80, clause 30.5; JCT 81, clause 30.4; IFC 84, clause 4.4; JCT 87, clause 4.8)** The requirements for the Employer to hold Retention in trust, and place it in a trust account, have been deleted. Such a requirement might be inappropriate, for example, if the Employer itself is a bank or financial institution of good standing. Alternatively, the Employer may be borrowing all the money in respect of a particular progress payment and may not therefore possess the relevant Retention. The Contractor should carefully consider in each case whether to take the relevant credit risk. Incidentally, the author has noticed on several occasions in practice that Contractors have failed to require Retention to be placed in a trust account when contractually entitled to make such a requirement, only to wake up to their rights when the Employer has become insolvent and such rights are valueless. Where Contractors have such a right they should invariably exercise it at the beginning of the Contract.

23. **Payment for Off-Site Materials or Goods (JCT 80, clause 30.3; JCT 81, Appendix 2; IFC 84, clause 4.2.1(c))** Discretionary provisions in respect of payment for off-site materials or goods have been deleted. In the case of JCT 87, the relevant provision is clause 4.22 of the Works Contract Conditions (Works Contract/2). If the Employer wishes, he may require that sub-clause to be deleted from Works Contracts under clause 8.2.1.1 of JCT 87, as amended by Form 16. Clause 30.3 of JCT 80 gives the Architect a discretionary power to include the value of off-site materials and goods in Interim Certificates, subject to a number of safeguards, such as passing of title to the Employer under clause 16. Despite the safeguards, the insolvency of the Contractor, or of the sub-contractor or supplier who actually has possession of the off-site materials and goods, usually leads to a shambles. Therefore, it is suggested that the Employer should only pay for materials and goods delivered to site or incorporated into the Works. As a matter of fact, the Employer can only ever be certain who owns materials and goods when they have been incorporated into the Works. By operation of law, they then vest in the freeholder, whoever owned them before. Even though goods and materials are delivered to site by the Contractor and paid for by the Employer, so that title **should** pass to the Employer under clause 16.1, the Employer can never be quite sure that they belong to him, while they are unfixed. Suppose, for example, that the Contractor had bought them in good faith from someone who had stolen them?

The true owner could recover them from the Employer and the Contractor would be liable to the Employer in damages. However, an Employer who pays for goods and materials delivered to site has as much assurance of title as anyone else who pays for goods delivered. Therefore, it is probably a counsel of perfection to suggest that payment should only take place in respect of materials and goods actually incorporated into the Works.

24. **Finality of Final Certificates (JCT 80, clause 30.9.2.2; JCT 81, clause 30.8.2; JCT 87, clause 1.14.2.2)** Clause 30.9.1 of JCT 80 renders the Final Certificate conclusive as to all important financial matters. By virtue of clause 30.9.2, mere lapse of time in arbitration proceedings without a formal step being taken could cause the Final Certificate to become conclusive. Clause 30.9.2 reads as follows:

'30.9.2 If any arbitration or other proceedings have been commenced by either party **before** the Final Certificate has been issued the Final Certificate shall have effect as conclusive evidence as provided in clause 30.9.1 after either:

.2.1 such proceedings have been concluded, whereupon the Final Certificate shall be subject to the terms of any award or judgment in or settlement of such proceedings, or

.2.2 a period of **12 months during which neither party has taken any further step** in such proceedings, whereupon the Final Certificate shall be subject to any terms agreed in partial settlement,

whichever shall be the earlier.'

It is suggested that this 'black hole' for each party (and most particularly for their respective solicitors) should be removed. It is perfectly possible for 12 months to go by without any 'further step' being taken, but without the parties intending to drop the matter.

25. **Fluctuations (JCT 80, clauses 38.4.8.1, 39.5.8.1 and 40.7.2.1; JCT 81, clauses 36.4.8.1, 37.5.8.1 and 38.6.2.1; IFC 84, Supplementary Conditions C4.8.1 and D13.1; MW 80, clause A4.4.4.2.1, Supplementary Memorandum)** The reason for the deletions of these sub-clauses has been mentioned earlier, in relation to extension of time for completion. Clause 38.4.8.1 of JCT 80 provides that clause 38.4.7, which effectively stops fluctuations if the Contractor is in unexcused delay, shall **not** be applied unless **'the printed text of clause 25 is unamended and forms part of the Conditions'**. Therefore, the

Contractor would continue to receive fluctuations during unexcused delay. As clause 25, in practice, will almost always have been amended by the deletion of clause 25.4.10, this is a most unreasonable pitfall for the Employer. It is also a particularly obscure pitfall. There is nothing in clause 25 to warn the draftsman that amending it could affect fluctuations. Moreover, clauses 38, 39 and 40 are actually published by the JCT as a separate document from JCT 80, and the effect of the relevant sub-clauses is by no means obvious, except when examined with considerable attention. In the case of JCT 87, the equivalent provisions are clauses 4A.4.8.1, 4B.5.8.1 and 4C7.3 of the Works Contract Conditions (Works Contract/2). If the Employer wishes, he may require the relevant sub-clauses of the Works Contract Conditions to be deleted under clause 8.2.1.1 of JCT 87, as amended by Form 16. See also clause 2.13.2 of JCT 87, as amended by Form 16, in respect of clause 2.10.10 of the Works Contract Conditions.

CHAPTER 3

COLLATERAL WARRANTIES

1. Why do we have collateral warranties at all? The reason is that, owing to recent decisions by the courts, it appears that the liability in tort for negligence of Contractors and consultants to third parties, other than their own Employers or clients, in respect of financial loss arising from construction defects is practically non-existent. See, for example, *D & F Estates Limited v Church Commissioners for England* [1989] AC177, HL, *Murphy v Brentwood District Council* [1990] 3 WLR 414, HL, and *Department of the Environment v Thomas Bates & Son Limited* [1990] 3 WLR 457, HL. Therefore, third parties will now usually demand collateral warranties in their favour from Contractors and consultants. In a conventional building project, where the Employer engages a consultant to design the Works and a Contractor to build them in accordance with that design, collateral warranties will usually be required from the Contractor and consultants concerned in favour of the funding institution and future purchasers and tenants. Warranties may be demanded from more than one consultant, e.g. Structural Engineers as well as Architects. In 'design and build' or 'turnkey' projects, where the Contractor is responsible for design as well as construction, the warranties required will differ, because it will be the Contractor, not the Employer, who will act as the client of the consultant. Therefore, the consultant should be requested to enter into warranties with the Employer, as well as with the funding institution, purchasers and tenants. Collateral warranties are much in need of standardization and practice differs very widely at the present time. The JCT publishes a number of documents to be used in cases where Nominated Sub-Contractors and the like assume direct contractual responsibilities **to the Employer**, e.g. JCT Agreement NSC/W, but these have no application to other parties such as funding institutions, purchasers and tenants.

2. The Forms given in this book try to deal with most aspects of collateral warranties which are likely to be encountered in practice. The clauses relating to collateral warranties to be given by Contractors (see Form 12) have to be rather complicated, but may be summarized as follows:

    — The Contractor is required to give collateral warranties to funding institutions and acquirers of interests in the Works. Their form is to be set out in Annexes. As the funding institution and acquirers may well not yet be identified, the Employer is authorized to make reasonable changes to the forms required. If the Contractor is a subsidiary, parent company guarantees of the collateral warranties are required.

– The Contractor is also required to obtain collateral warranties in favour of the Employer, funding institutions and acquirers from certain sub-contractors and suppliers, with parent company guarantees in respect of those sub-contractors and suppliers.

– If a sub-contractor or supplier will not give the prescribed collateral warranty, or will only give it in a modified form, the Employer may require that the sub-contractor or supplier shall not be used, or may accept that the sub-contractor or supplier need not give the collateral warranty, or may give it in a modified form.

– In the case of a Nominated Sub-Contractor or Nominated Supplier under JCT 80, the Contractor cannot be required to use such a sub-contractor or supplier who will not give collateral warranties acceptable to the Employer.

– Provision is made for the eventuality that the Contractor may not complete the execution of all the Works.

As mentioned in Chapter 1, all the collateral warranties suggested in this book are based upon BPF Form CoWa/F, Second Edition 1990 (Appendix B).

3.   The clauses relating to collateral warranties to be given by consultants (see Form 1) are simpler than those included in building contracts, as there is not the same need to deal with sub-contractors.

4.   In practice, the clauses may well be watered down by, for example, requiring that only purchasers or tenants of the **whole** property shall receive collateral warranties, or by requiring sub-contractors or suppliers to give collateral warranties only to the Employer.

5.   There is interminable argument about collateral warranties. The author's view is that potential warrantors are in a poor position to argue rationally that they should not be liable (to the same extent to which they are liable to their own clients or Employers) to funding institutions and acquirers of interests in the Works. If the client or Employer himself occupies the Works, consultants and Contractors who have engaged in relevant contracts with the Employer are liable to him for breach of those contracts for 6 years, or 12 years if (as they should have been) the relevant contracts were entered into as Deeds. Therefore, the relevant consultants and Contractors may be effectively released from liability prematurely if the Works are sold, or let on full repairing terms, without their giving suitable collateral warranties to the relevant third parties.

6.   As mentioned below in relation to letters of intent, the best way to ensure that

consultants and Contractors will give acceptable collateral warranties is to write such documents into their respective contracts at the outset; and to require those contracts to be formally entered into before the client or Employer makes any legal commitment to the consultants and Contractors concerned, or parts with any funds.

7. It is frequently argued that the warrantor's liability should be limited to the cost of physical repairs, leaving consequential loss, such as the cost of alternative premises, to fall on the warrantee. For example, see clause 1 of BPF Form CoWa/P&T (Appendix D). However, a warrantor's liability to his own client or Employer, and the warrantor's professional indemnity insurance cover, will not usually be limited in the like way.

8. Warrantors and their professional indemnity insurers now often ask for a 'net contribution' clause. This is intended to avoid the consequences of the insolvency of a party who is jointly liable with the warrantor to the warrantee in respect of the same matter. This would normally mean that the solvent warrantor would have to pay all the damages to which the warrantee is entitled, including the insolvent's share. A 'net contribution' clause will usually be to the effect that the solvent warrantor will only be liable for his just share of the damages, leaving the loss of the insolvent party's share of the damages to fall on the warrantee. See, for example, clause 1 of BPF Form CoWa/F, Third Edition 1992, and of Form CoWa/P&T (Appendices C and D).

9. Employers should consider, with their insurance advisers, latent defects insurance (normally for ten years), in addition to contractual obligations and collateral warranties. In the case of dwellings, the National House Building Council ('NHBC') scheme is, of course, available, as well as other more recently established similar schemes.

CHAPTER 4

LETTERS OF INTENT

1. It very frequently occurs in practice that Contractors and consultants will start work under 'letters of intent', before entering into a formal contract. This can be for many reasons, but perhaps the two most common are lack of final agreement upon detailed prices and other terms and conditions, and the Employer's desire to start work on site at an early date. Lack of final agreement upon details will very often relate to collateral warranties, the agreement of which may involve third party funding institutions and their professional advisers, as well as the potential warrantors' professional indemnity insurers.

2. Letters of intent should be avoided whenever possible but, if they are absolutely necessary, should always be drafted with legal advice. They are usually preliminary contracts enabling work to be executed and paid for an an interim basis and are frequent sources of very serious trouble, especially for the Employer. For example, see *British Steel Corporation v Cleveland Bridge & Engineering Company Limited* (1981) 24 BLR 94 and *Kitsons Insulation Contractors Limited v Balfour Beatty Building Limited* (8-CLD-05-04).

3. Letters of intent may assume an infinite variety of forms, and the factors which need to be borne in mind include the following:

    - issuing a letter of intent usually weakens the Employer's bargaining position, because severance of the relationship will mean delay in the Works and sometimes double costs, such as mobilization costs;
    - the unintentional conclusion of a contract for the entire Works by means of the letter of intent should be avoided;
    - the basis, method and frequency of payment for work executed pursuant to the letter of intent should be clearly determined;
    - the Employer may wish to impose a financial ceiling on his liability for payment under the letter of intent;
    - insurance, particularly insurance of the Works, should be considered;
    - the consequences of failure to agree on a formal contract should be clear (usually, this will involve payment for work done, and a severance of the relationship); and
    - the procedure for terminating the letter of intent should be specified (preferably, the Employer should be able to terminate at will).

4. The period of time spent working under a letter of intent should be minimized, or insuperable problems may arise. For example, what will happen if a substantial portion of the Works is executed pursuant to a letter of intent, but agreement cannot be reached on collateral warranties? If the Employer then engages alternative Contractors or consultants, there will be a lack of collateral warranties in respect of the portion of the Works already executed.

CHAPTER 5

CONSTRUCTION MANAGEMENT AND THE JCT CONTRACTS*

Synopsis

This paper outlines the contractual background to construction management and other forms of construction procurement, and the Employer's existing choices under the JCT Conditions of Contract, particularly management contracting. The provisions of the JCT Management Contract 1987 relieving the Management Contractor from liability and dealing with determination of the Management Contractor's employment are considered, together with the Management Contractor's ambiguous position. Construction management as an alternative to management contracting is then explored, including the possible adaptation of JCT 80 for construction management. It is concluded that, in a Project which is suitable either for construction management or management contracting, there seem to be patent advantages to the Employer in choosing construction management. The views expressed in this paper are those of the writer, and not necessarily those of Simmons & Simmons.

Glossary

Acquirer: A purchaser, tenant, mortgagee or other person acquiring an interest in the Project.

AIA: The American Institute of Architects.

AIA Handbook: 'The Architect's Handbook of Professional Practice' (11th edn), published by the AIA.

Barber: Peter W. B. Barber 'Management Contracting: Low Risk for Contractor – Low Recovery by Employer': the Society of Construction Law Prize Paper, 1988.

Fund: Any funding institution financing the Employer in relation to the Project.

* This chapter consists of a paper written by the author and submitted in the 1991 Alfred Hudson Prize Competition (organized annually by the Society of Construction Law). It was awarded joint third prize and was first published by the Society of Construction Law in 1992. It was also published in 'The International Construction Law Review' in October 1992 ([1992] ICLR 476).

GC Works/1:	Department of the Environment 'Form GC/Works/1: General Conditions of Contract – Edition 3: Standard Form for a Lump Sum Contract, With Quantities', published December 1989.
IFC 84:	JCT Intermediate Form of Building Contract, 1984 Edition.
JCT:	The Joint Contracts Tribunal for the Standard Form of Building Contract, whose constituent bodies are:

- Royal Institute of British Architects;
- Building Employers Confederation;
- Royal Institution of Chartered Surveyors;
- Association of County Councils;
- Association of Metropolitan Authorities;
- Association of District Councils;
- Confederation of Associations of Specialist Engineering Contractors;
- Federation of Associations of Specialists and Sub-Contractors;
- Association of Consulting Engineers;
- British Property Federation;
- Scottish Building Contract Committee.

JCT 80✳:	JCT Standard Form of Building Contract, 1980 Edition, incorporating Amendments 1–9, Private With Quantities.
JCT 81:	JCT Standard Form of Building Contract With Contractor's Design, 1981 Edition.
JCT 87:	JCT Standard Form of Management Contract, 1987 Edition.
MC/1:	JCT Practice Note MC/1: Management Contracts under the JCT Documentation.
MC/2:	JCT Practice Note MC/2: Commentaries on the Management Contract Documentation 1987.

✳ Amendments 10 and 11 have since been issued.

MW 89:	JCT Agreement for Minor Building Works.
NSC/2✳:	JCT Standard Form of Employer/Nominated Sub-Contractor Agreement, for use with JCT 80.
NSC/2a✳:	Agreement NSC/2 adapted for use when Tender NSC/1 has not been used, for use with JCT 80.
Rowe & Maw:	Rowe & Maw 'The JCT Management Contract' (Sweet & Maxwell, 1989).
Trade Contract, Trade Contractor:	A direct contract or contractor with the Employer **under a construction management Project**.
Works Contract, Works Contractor:	A sub-contract or sub-contractor **under a Management Contract.**
Works Contract/2:	JCT Works Contract Conditions, for use with JCT 87.
Works Contract/3:	JCT Standard Form of Employer/Works Contractor Agreement, for use with JCT 87.
Works Package:	The Works included in each Trade or Works Contract.

Text

Contractual Background

1. This paper is concerned with 'construction management', a contractual structure under which:

    – there is no single Main Contractor;
    – the Employer enters into several direct Trade Contracts for Works Packages comprising portions of the Project; and
    – the Employer instructs a Construction Manager to run the Project.

2. The Construction Manager may, but certainly need not be, a contractor. His role may be fulfilled by professional consultants, such as Architects, Quantity Surveyors or Engineers, provided that they possess the necessary construction management experience. The Construction Manager may be an external

✳Since replaced by NSC/W.

consultant, or an employee of the Employer.

3. The function of the Construction Manager is to co-ordinate the work of the several Trade Contractors, and of the Employer's other consultants.

4. The paper also reviews the JCT Contracts in relation to construction management, particularly in order:

    − to compare the legal side of management contracting and construction management; and
    − to suggest how JCT documents might be adapted for use in construction management projects.

5. In order to set the scene, certain Figures are included in this paper. **Figure 1** shows the principal contractual relations under construction management. The Employer enters into terms of engagement separately with his Construction Manager (unless the Construction Manager is 'in-house') and also with his Architect, Quantity Surveyor, Structural Engineer and Building Services Engineer. Of course, some (or even all) of these functions may be combined in one firm. There may be additional consultants, such as Landscape Architects, in certain Projects, and sometimes one or more of the usual consultancy functions may not be required. However, the writer has tried to show a typical range of consultants.

6. The Employer also enters into separate direct Trade Contracts for the various Works Packages with the several Trade Contractors. Five are shown, but there may be forty or fifty.

7. The terms 'Trade Contracts' and 'Trade Contractors' are used in relation to direct Contracts with the Employer in a construction management Project, in contrast to the terms 'Works Contracts' and 'Works Contractors', which are used in relation to sub-contracts under management contracting. The fundamental difference is that Trade Contracts are direct Contracts with the Employer, while Works Contracts are sub-contracts from the Management Contractor.

8. Each Trade Contractor will usually have his own Sub-Contractors. Under management or other forms of contracting, they would be Sub-Sub-Contractors.

9. It is a fundamental difference between construction management and other forms of contracting that one contractual layer − the single Main Contract for the Project − does not exist.

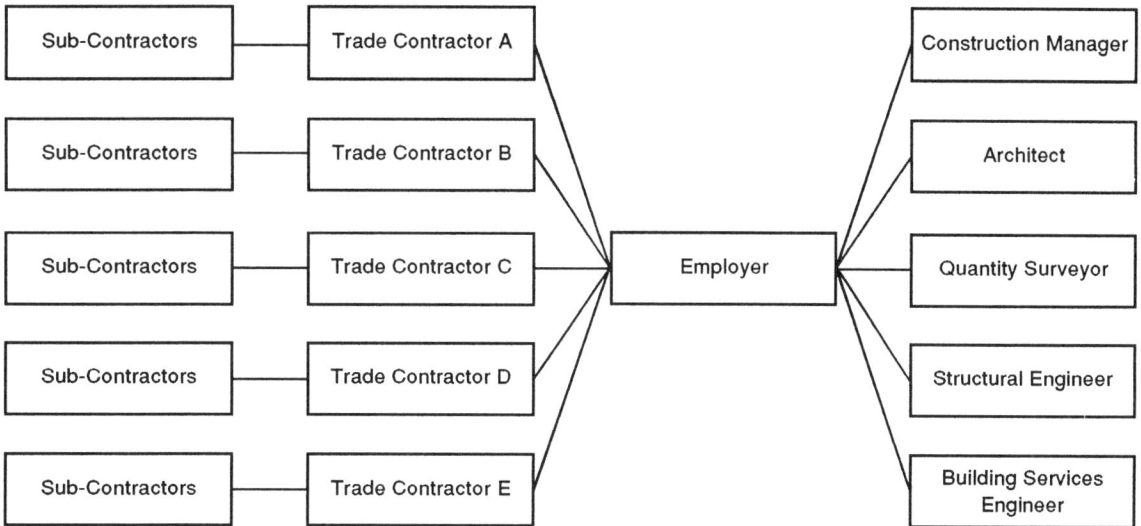

Figure 1 Principal Contractual Relations under Construction Management

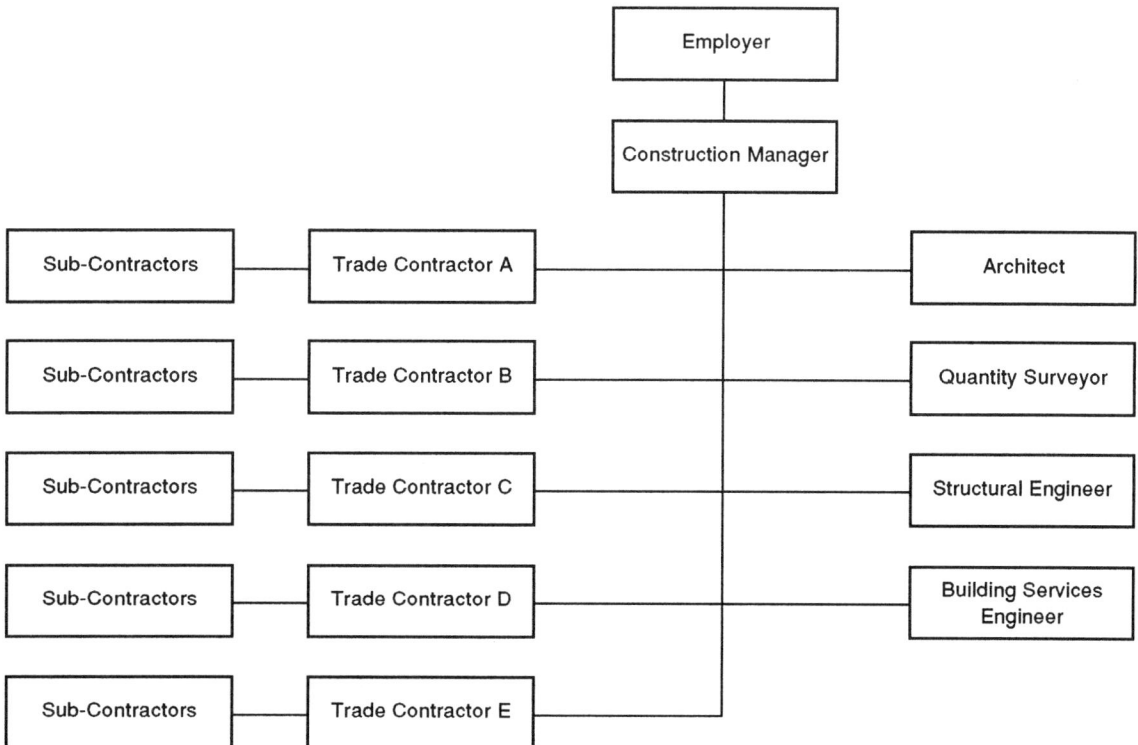

Figure 2 Organization under Construction Management

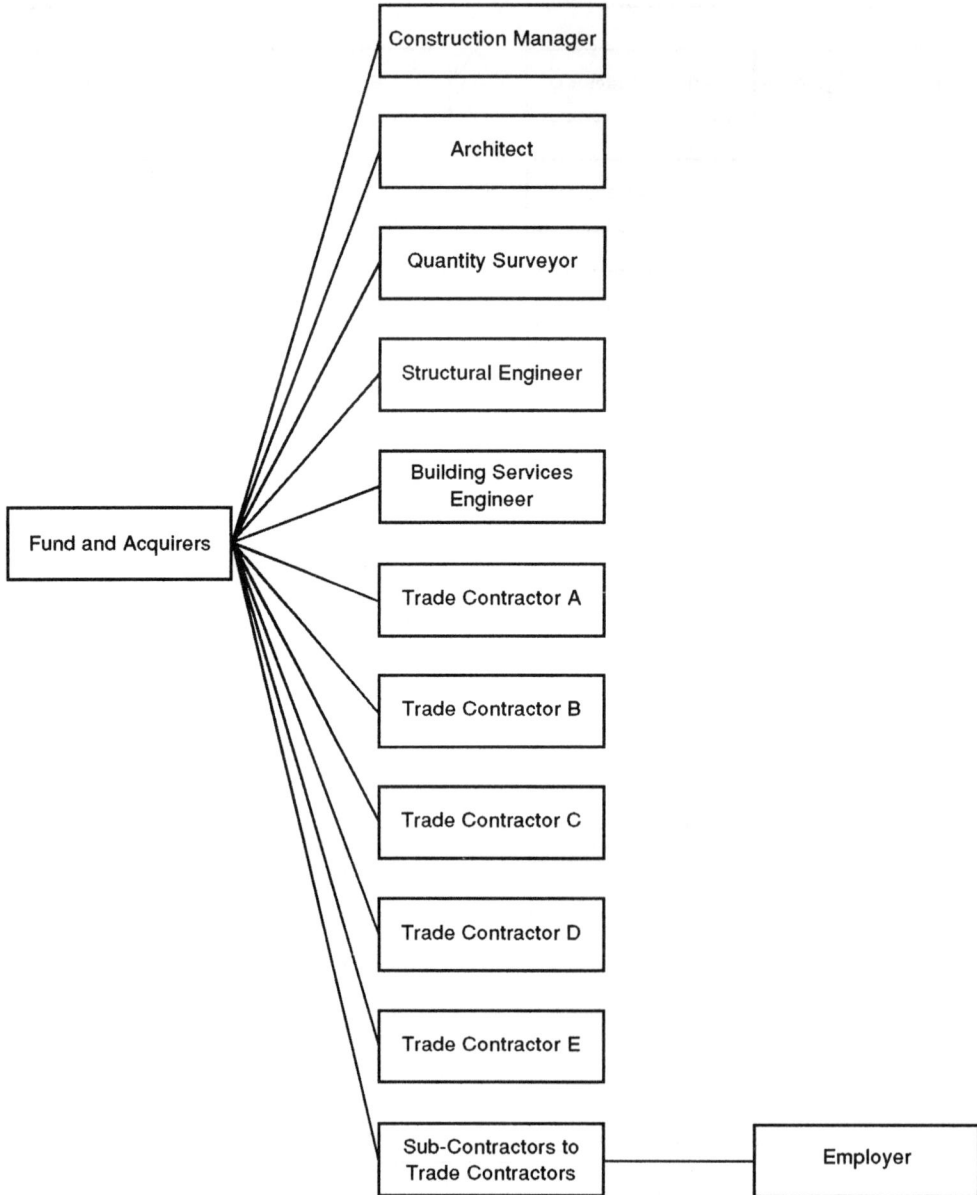

Figure 3 Collateral Warranties under Construction Management

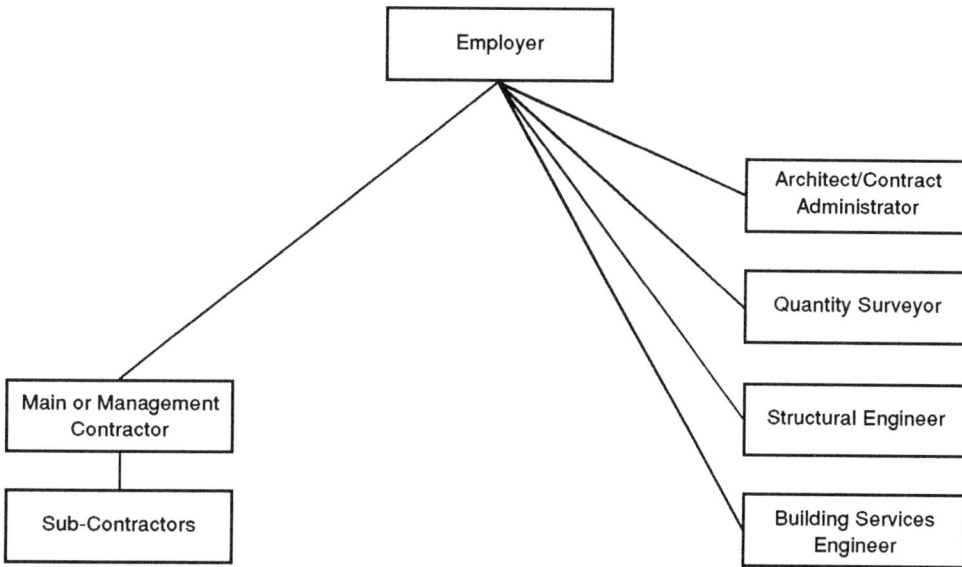

Figure 4 Principal Contractual Relations under Lump Sum and Management Contracting

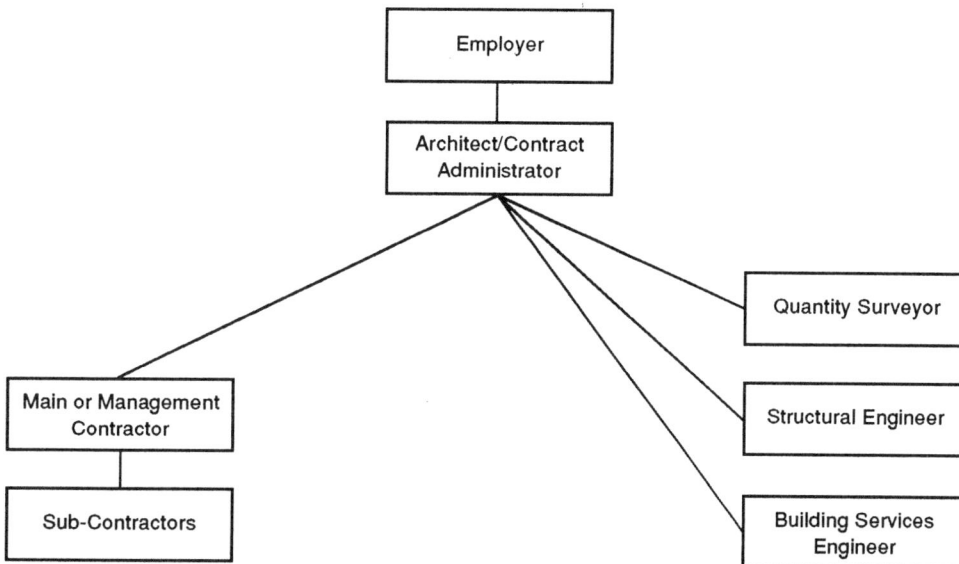

Figure 5 Organization under Lump Sum and Management Contracting

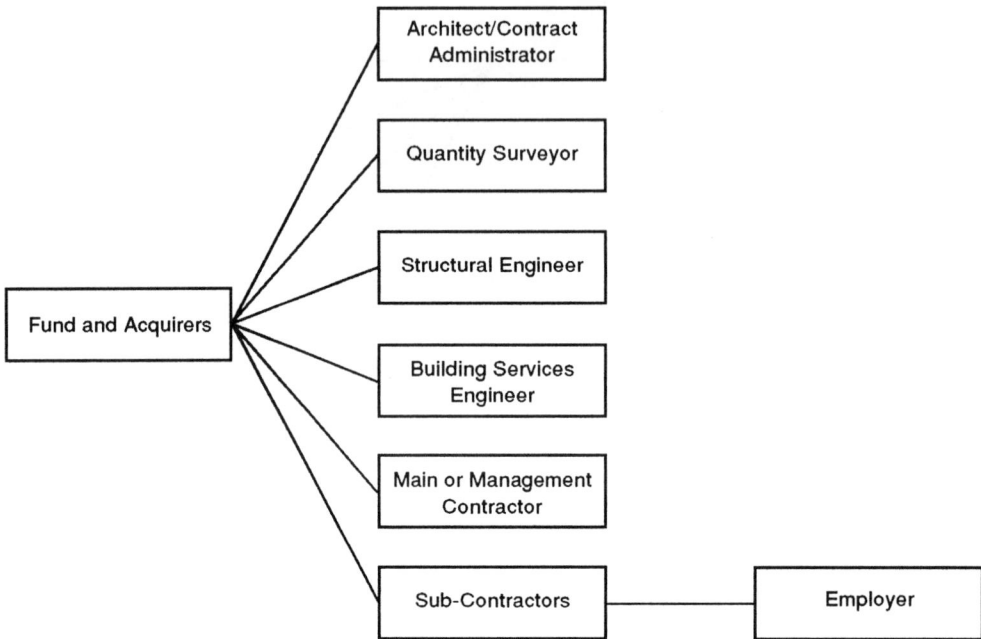

Figure 6 Collateral Warranties under Lump Sum and Management Contracting

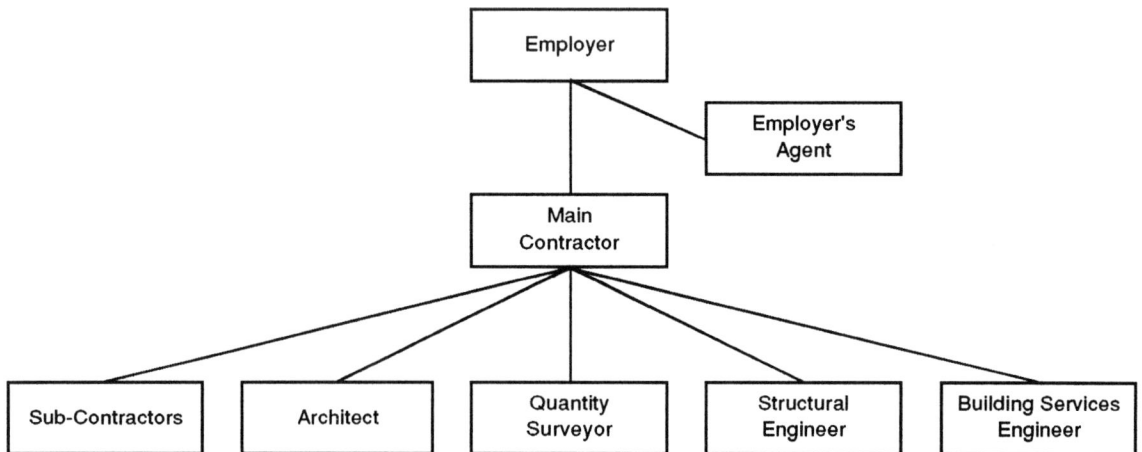

Figure 7 Principal Contractual Relations under 'Design and Build'

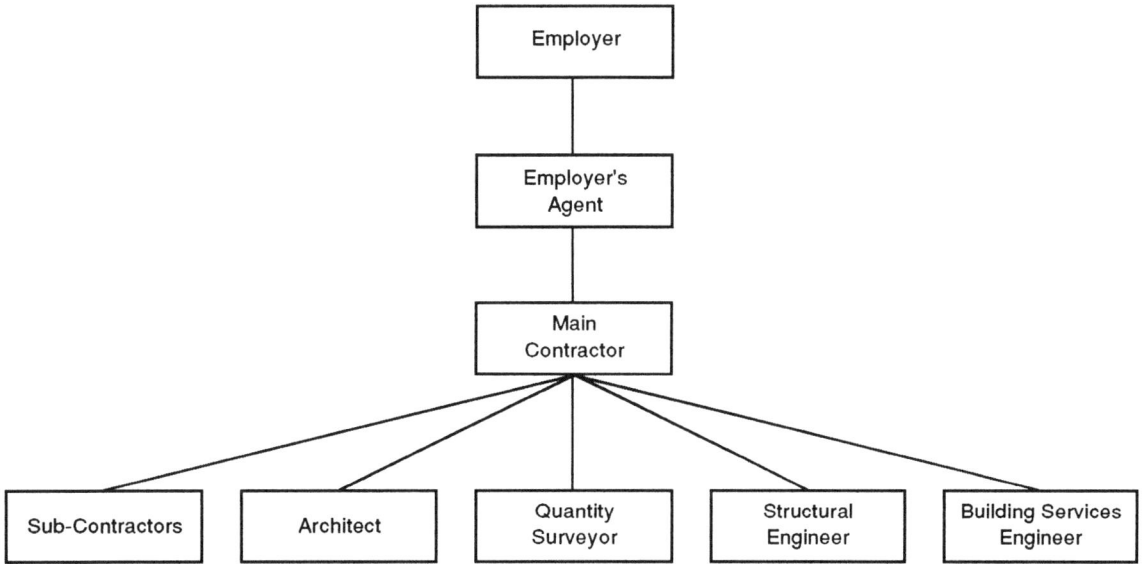

Figure 8 Organization under 'Design and Build'

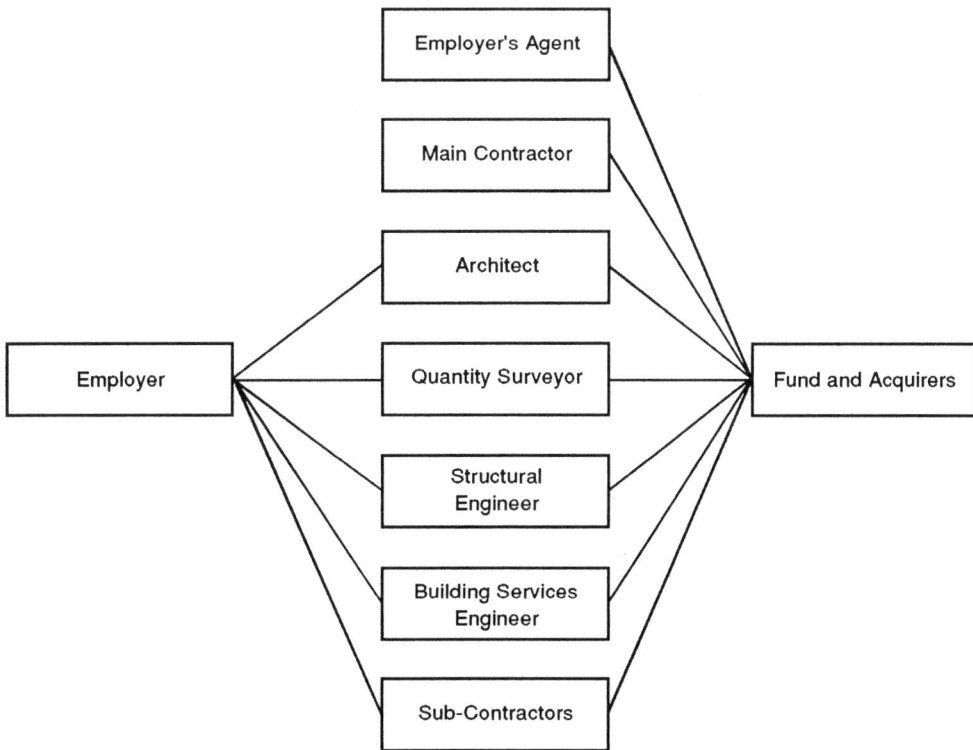

Figure 9 Collateral Warranties under 'Design and Build'

10. **Figure 2** shows organization under construction management – that is, the control and communication structure of the Project. The Employer receives information from, and issues instructions to, the Construction Manager, who is the clearing-house. The Construction Manager receives information from, and issues instructions on behalf of the Employer to, the other consultants and the Trade Contractors, and co-ordinates the activities of all of them.

11. **Figure 3** shows collateral warranties under construction management. Collateral warranties may be required from the Construction Manager, the Employer's other consultants, the Trade Contractors and perhaps certain Sub-Contractors, in favour of the Fund and Acquirers. Collateral warranties may also be required from certain Sub-Contractors in favour of the Employer. Collateral warranties will usually be required only from major Sub-Contractors, especially those with design responsibilities in respect of their Sub-Contract Works.

12. **Figure 4** shows the principal contractual relations under both lump sum and management contracting. The Employer enters into terms of engagement with each of the consultants, usually led by an Architect or Contract Administrator, and into one Main or Management Contract for the Project with the Main or Management Contractor, who will usually sub-contract the great bulk of the Project to Works Contractors.

13. **Figure 5** shows organization under lump sum and management contracting. The Employer receives information from, and issues instructions to, the Architect or Contract Administrator, who will generally co-ordinate the other consultants' work. The Main or Management Contract will usually specify that only the Architect or Contract Administrator may issue instructions on behalf of the Employer to the Main or Management Contractor. The organization and management of the construction process itself (a major part of the Construction Manager's duties under construction management) is the duty of the Main or Management Contractor and not of the Architect or Contract Administrator.

14. **Figure 6** shows collateral warranties under lump sum and management contracting. Collateral warranties may be required from the Architect or Contract Administrator, the Employer's other consultants, the Main or Management Contractor and certain Works Contractors or Sub-Contractors, in favour of the Fund and Acquirers, and from certain Works Contractors or Sub-Contractors in favour of the Employer. The latter will usually include NSC/2✱ or NSC/2a✱ from Nominated Sub-Contractors where JCT 80 is being used as the Main Contract, or Works Contract/3 where JCT 87 is being used as the

✱Since replaced by NSC/W.

Main Contract. If numerous collateral warranties by Works Contractors or Sub-Contractors in favour of the Fund or Acquirers are demanded, there may be as many collateral warranties under lump sum or management contracting as under construction management.

15. **Figure 7** shows the principal contractual relations under 'design and build'. The Architect and other consultants act as 'sub-contractors' to the Main Contractor, who (unless they are 'in-house' to the Main Contractor) will enter into terms of engagement with the consultants, and into whatever Sub-Contracts are required in respect of the Works. The Employer enters into the Main Contract and into terms of engagement with his Employer's Agent (the term used in JCT 81), unless the Employer's Agent is 'in-house' to the Employer.

16. **Figure 8** shows organization under 'design and build', which is fairly self-explanatory.

17. **Figure 9** shows collateral warranties under 'design and build'. Collateral warranties may be required from the Architect and the Main Contractor's other consultants in favour of the Employer, as the Employer has no terms of engagement with them. The Employer will also usually require collateral warranties from certain Sub-Contractors, as in other forms of contracting. The Main Contractor, the Architect, the Main Contractor's other consultants and certain Sub-Contractors will usually also be required to give collateral warranties in favour of the Fund and Acquirers. The Employer's Agent may also be required to give collateral warranties in favour of the Fund and Acquirers, if he fulfils a sufficiently important role – such as formulating, or advising upon, the Employer's Requirements included in the 'design and build' Contract.

Employers' Choices under JCT Conditions of Contract

18. The JCT conditions of contract are dominant in UK construction projects, with the exception of Central Government projects where GC Works/1 or its minor works equivalent GC Works/2 may be used. However, in UK civil engineering projects, the Institution of Civil Engineers ('ICE') Conditions of Contract are dominant, with the same exceptions for Central Government projects.

19. The JCT conditions are suitable for three principal types of construction Project:

     – lump sum contracts, where the Employer's Architect designs the Works (JCT 80, IFC 84 and MW 89);

- lump sum contracts, where the Contractor designs the Works (JCT 81); and
- management contracting (JCT 87).

20. The common law is hard upon a lump sum Contractor. All unforeseeable risks and costs, such as exceptionally adverse weather conditions and unexpected inflation, are generally laid upon his shoulders and he is strictly responsible to the Employer for all his sub-contractors, including those imposed upon him (or 'nominated') by the Employer under the terms of the relevant Main Contract. Over the years, Contractors have engaged in a long march in order to improve their position, with considerable success in relation to the JCT conditions of contract.

21. As a result, JCT 80 contains numerous provisions protecting the Contractor, notably with regard to extensions of time for completion of the Works, Nominated Sub-Contractors and inflation. For example, clause 25 (Damages for non-completion) of JCT 80 contains a long list of 'Relevant Events' which entitle the Contractor to an extension of time for completion of the Works. Clause 25.4.10 includes as Relevant Events:

'25.4.10.1 the Contractor's inability for reasons beyond his control and which he could not reasonably have foreseen .... to secure such labour as is essential to the proper carrying out of the Works; or

.10.2 the Contractor's inability for reasons beyond his control and which he could not reasonably have foreseen .... to secure such goods or materials as are essential to the proper carrying out of the Works.'

22. Not surprisingly, a well-advised Employer will normally delete clause 25.4.10 and, consequently, the insidious pitfalls in clauses 38.4.8.1, 39.5.81 and 40.7.2.1, which permit price fluctuations to continue even when the Contractor is in unexcused delay if the **printed text** of clause 25 is altered.

23. With regard to **Nominated** Sub-Contractors, the Main Contractor is very largely relieved of responsibility for their performance, except for their supply of workmanship, materials and goods. By clause 25.4.7 of JCT 80, delay on the part of Nominated Sub-Contractors is a Relevant Event entitling the Main Contractor to an extension of time for completion of the Works. By clause 35.21, the Main Contractor is effectively relieved of responsibility for design by Nominated Sub-Contractors, which can be very important in relation to such matters as lifts and cladding. Both delay and defective design by Nominated Sub-Contractors are matters directly between the Employer and the Nominated

Sub-Contractor, by virtue of NSC/2✻ or NSC/2a✻. Any adverse financial effect of a Nominated Sub-Contractor's insolvency is cast upon the Employer by clause 35.24.

24. The result of the above provisions in respect of Nominated Sub-Contractors has been to make many well-advised Employers avoid the nomination of Sub-Contractors wherever possible. This may be done by the use of clause 19.3, under which the Main Contractor is given a list of not less than three prospective Sub-Contractors in respect of certain Sub-Contract Works, and is entitled to choose between them, in which case they will be not be Nominated Sub-Contractors. Alternatively, where the Employer actually wishes to select the Sub-Contractor, but without 'nominating' him, all sorts of concepts, such as 'Specialist Sub-Traders' have been invented. However, the simplest course is to amend JCT 80 (for example, by the deletion of clause 25.4.7 and the amendment of clauses 35.21 and 35.24) in order to remove the disadvantages to the Employer which would otherwise flow from the nomination of Sub-Contractors. The widespread unacceptability of the Nominated Sub-Contractor provisions is one of the major difficulties with JCT 80 in practice.

25. With regard to inflation, the Contractor may also be protected, to a greater or lesser extent, by contractual fluctuations in the Contract Sum, under one of the alternative clauses 38 (Contributions, levy and tax fluctuations), 39 (Labour and materials cost and tax fluctuations) or 40 (Use of price adjustment formulae).

26. The initial Contract Sum under JCT 80 may also be rather 'flexible' because, for one reason or another, the Contract is often awarded when the Works have been only partly designed. The initial Contract Sum may, therefore, include a large proportion of Provisional or Prime Cost Sum, and the Contractor may subsequently have wide scope for claims (for example, under clause 26.2.1) resulting from the design being completed while the Works are actually in progress.

27. Therefore, Employers often have surprises in store when awarding Contracts on JCT 80 terms.

28. Nevertheless, if an Employer wishes to know in advance with reasonable precision what the Works will cost and what they will consist of when completed, he is probably best advised:

✻ Since replaced by NSC/W

- to have the Works substantially designed, before awarding the Contract;
- to award the Contract on a lump sum basis, using JCT 80, or IFC 84 or MW 89 for smaller Projects, suitably amended; and
- to avoid both subsequent changes of mind and the use of Nominated Sub-Contractors.

29. If the Employer is prepared to accept that he will not completely control the design of the Works, and desires a quick start on Site, he might wish to use 'design and build' under JCT 81, but such an Employer should realise that the Main Contractor is entitled to design and build the Works within the parameters set by the Employer's Requirements and Contractor's Proposals included in the Contract. As JCT 81 is a 'lump sum' Contract, like JCT 80, the Contractor has an inevitable incentive to build down to the minimum acceptable under the Employer's Requirements and Contractor's Proposals.

30. The fundamental difference for financial risk purposes is between 'lump sum' contracting and 'cost plus' contracting. If an adventurous Employer decides to depart from lump sum contracting altogether and pay the prime cost of the Works plus a fee for management of the Project, he has a choice between construction management and management contracting. The main difference between the two consists in whether to have a Main or Management Contractor, rather than a Construction Manager. In either event, the Employer will be taking upon himself a great deal of what would have been the Main Contractor's risk and responsibility under a lump sum Contract, but will have the opportunity to make at least some of the Main Contractor's profit. JCT 87 has been published by the JCT for management contracting, but the JCT has published no Contract for construction management.

31. MC/1, page 17, states that:

'Suitable conditions for the use of the JCT [Management Contract] documents (though not all are necessary for a successful Management Contract) are:
(a) the Employer wishes the design to be carried out by an independent Architect and design team;
(b) there is a need for early completion;
(c) the project is fairly large;
(d) the project requirements are complex;
(e) the project entails or might entail changing the Employer's requirements during the building period;
(f) the Employer while requiring early completion wants the maximum possible competition in respect of the price for the building works.'

32. Such conditions are equally suitable for the use of construction management.

JCT 87: Management Contractor's Relief from Liability

33. JCT 87 envisages that all the Works Packages included in the Project (apart from certain site facilities and services to be provided by the Management Contractor under clause 1.5.4 and the Fifth Schedule) will be sub-contracted by the Management Contractor to Works Contractors. The responsibility of the Management Contractor for Works Contractors is intended to be restricted by clauses 1.7 and 3.21 of JCT 87, which read as follows:

'1.7   **Subject to clause 3.21** the Management Contractor shall be fully liable to the Employer for any breach of the terms of this Contract including any breach occasioned by the breach by any Works Contractor of his obligations under the relevant Works Contract.'

'3.21   **Notwithstanding anything contained elsewhere in this Contract** the following provisions shall apply in respect of any breach of, or non-compliance with, a Works Contract by a Works Contractor (which shall be deemed to include a determination of the employment of a Works Contractor under .... the Works Contract Conditions and also the engagement, as a result of such breach or non-compliance, of other persons to carry out part or the whole of the Works Contract Works in accordance with .... the Works Contract Conditions):

3.21.1 The Management Contractor shall in consultation with the Architect/ the Contract Administrator and the Employer take all necessary steps

.1 to operate the terms of the Works Contract for dealing with such breach or non-compliance, **including enforcement through arbitration or litigation if necessary,** to obtain any amount due to the Management Contractor including therein any amount for which the Management Contractor is liable to the Employer under clause 1.7, as a result of the breach or non-compliance by the Works Contractor; and

.2 to secure the satisfactory completion of the Project including the engagement for that purpose of a further Works Contractor if such engagement is in accordance with the terms of the Works Contract with the Works Contractor who has failed to comply with the Works Contract or is in breach or is necessary because the employment of the Works Contractor under that Works Contract has been determined because of a breach or non-compliance; and

.3 to meet any claims properly made under the Works Contract Conditions, by Works Contractors, other than the Works Contractor who is in breach or who has failed to comply with the Works Contract, in respect of the consequences to them of such breach or non-compliance.

.2 The Employer shall

.1 pay to the Management Contractor in accordance with Section 4 and the Second Schedule all amounts properly incurred by the Management Contractor in fulfilling the obligations set out in clauses 3.21.1.1 and 3.21.1.2 but subject to the right of recovery by the Employer referred to in clause 3.21.2.3; and

.2 keep an account of any liquidated and ascertained damages due, but not deducted or recovered .... because the Completion Date has been exceeded by reason of the breach or non-compliance by a Works Contractor but shall not, except to the extent provided in clause 3.21.2.3, recover such damages from the Management Contractor;

.3 be entitled to recover from the Management Contractor all amounts paid or credited to the Management Contractor under clause 3.21.2.1 and where relevant the amount of liquidated and ascertained damages referred to in clause 3.21.2.2 **but only to the extent that such amounts have been recovered by the Management Contractor from the Works Contractor who is in breach or who has failed to comply with the Works Contract.**

.3 **In respect of the claims properly made by Works Contractors as referred to in clause 3.21.1.3** the Management Contractor shall be entitled to deduct from amounts in respect of the Works Contractor who is in breach or who has failed to comply with the Works Contract .... the amount of such claims which he has paid or is liable to pay to such Works Contractors together with any costs that he has incurred due to the breach or non-compliance. To the extent that the Management Contractor is not reimbursed by such deduction he shall seek to recover any shortfall in that reimbursement from the Works Contractor who is in breach or who has failed to comply with the Works Contract, through arbitration or litigation if necessary. If, despite compliance by the Management Contractor with the terms of clause 3.21.3, the Management Contractor is not fully reimbursed **then the Employer shall pay to the Management Contractor the amount of that shortfall in reimbursement.'**

34. With the intention of passing on the Management Contractor's liability to Works Contractors, and avoiding 'no loss' defences by Works Contractors, clause 1.6 of Works Contract/2 provides as follows:

'1.6.1   The Works Contractor shall be fully liable to the Management Contractor for any breach of the terms of the Works Contract. Such liability shall include, but shall not be limited to, any liability which the Management Contractor may incur to the Employer under or for breach of the Management Contract by reason of the negligence, act, omission or default of the Works Contractor.

1.6.2   The Works Contractor, having notice of the terms of the Management Contract ..., **undertakes not to contend, whether in proceedings or otherwise,** that the Management Contractor has suffered or incurred no damage, loss or expense or that his liability to the Management Contractor should be in any way reduced or extinguished by reason of clause 3.21 .... of the Management Contract Conditions.'

35. Works Contractors should also be required to enter into Works Contract/3 with the Employer, clauses 1 and 2 of which read as follows:

'1   The Works Contractor warrants that he has exercised, and will exercise, all reasonable skill and care in

.1 the design of the Works insofar as the Works have been or will be designed by the Works Contractor; and

.2 the selection of materials and goods for the Works insofar as such materials and goods have been or will be selected by the Works Contractor; and

.3 the satisfaction of any performance specification or requirement insofar as such performance specification or requirement is included or referred to in the description of the Works in the documents annexed to the Invitation to Tender ..., and which are included in the 'Numbered Documents' for the Works Contract as listed in the Works Contract Articles of Agreement ....

Nothing in Clause 1 shall be construed so as to affect the obligations of the Works Contractor under the Works Contract in regard to the supply under the Works Contract of workmanship, materials and goods.

2   The Works Contractor shall so supply the Architect/the Contract Administrator with information (including drawings) in accordance with any agreed programme or at such time as the Architect/the Contract Administrator may reasonably require so that the Architect/the Contract Administrator will not be delayed in issuing the necessary instructions and/or drawings and/or other documents to the Management Contractor in accordance with the Management Contract.'

36. Supposing that the legal result of these provisions is as apparently intended, how acceptable is the result from the Employer's point of view?

37. In the event of a defect in a Works Contractor's Works, it is very likely that the Employer will have to pursue both the Management Contractor and the Works Contractor, in view of his dual remedies under JCT 87 and Works Contract/3 and the common difficulty or impossibility in determining whether the cause of a defect is design or workmanship. Therefore, the Employer under JCT 87 will have the same multi-party problems as an Employer under JCT 80 who has problems with Nominated Sub-Contractors. The difference is that the Management Contractor is relieved of liability in respect of Works Contractors under clause 3.21 of JCT 87 even more effectively than the Main Contractor under JCT 80 is relieved of liability in respect of Nominated Sub-Contractors. In practice, the Employer will also often have to bring in the Architect, Structural Engineer etc. as parties, just as in disputes relating to JCT 80.

38. The Employer under JCT 87 will have the added complication that the Management Contractor may contend that, instead of the Management Contractor proceeding against the Works Contractors under clause 3.21.1, it would be preferable for the Employer to proceed against the Works Contractor under Works Contract/3.

39. The Management Contractor, in conducting disputes, litigation or arbitration under clause 3.21, has his costs and expenses paid by the Employer and the Prime Cost of his site employees will be reimbursed under the Second Schedule to JCT 87. However, in the absence of unusual special provisions entitling him to extra Management Fee, he has a great incentive to minimize his time and effort, and has little or no interest in the outcome. It will usually be in the Management Contractor's interests that the Employer's claims shall

fade away, especially after Practical Completion. Therefore, the Management Contractor may easily in practice become the Works Contractor's defender against the Employer, just as if the Management Contractor were a lump sum Contractor under JCT 80 or JCT 81.

40. There will sometimes be instances where the Works Contractor may be liable to the Management Contractor for damages which the Management Contractor has himself suffered and which the Management Contractor is not liable to pass on to the Employer, but such damages are of little or no interest to the Employer. Nevertheless, it appears that the Employer will have to pay the costs and expenses incurred by the Management Contractor in enforcing such claims against the Works Contractor.

41. Therefore, the Employer is placed in the position of having to pay for disputes, litigation or arbitration between the Management Contractor and Works Contractors, over which he has little effective control and in which his 'champion', the Management Contractor, may have little interest except in minimizing his own time and effort. If he disagrees with the Management Contractor's conduct of such litigation or arbitration, he has the theoretical remedy of damages against the Management Contractor, if he can prove breach of the Management Contract by reason of the Management Contractor's mismanagement. However, in the absence of flagrant misconduct or neglect by the Management Contractor, this may well be an illusory remedy in practice.

42. In these circumstances, one might well ask what useful purpose the Management Contractor is fulfilling for the Employer. Would it not be much better from the Employer's point of view to have these matters entirely in his own hands, by using several direct Trade Contracts and a Construction Manager, or anyone else he pleases, as his agent?

43. It has also been very seriously doubted whether the JCT 87 and Works Contract/2 provisions quoted above will operate as intended. See Rowe & Maw and Barber, amongst others. It is not practicable here to go into much detail, but it may be that there are great surprises in store for users of the JCT 87 documentation, when cases are litigated. For example:

   - both writers (Barber, paragraph 6.9; Rowe & Maw, D-35 and N-17) suggest that clause 1.6.2 of Works Contract/2 may not work at all, because the Management Contractor might, by reason of clause 3.21 of JCT 87, never be able to prove loss, notwithstanding that the Works Contractor, under clause 1.6.2 of Works Contract/2 is obliged 'not to contend' that the Management Contractor has suffered no loss;

— Barber (paragraph 10.2) suggests that because 'Clause 3.21 itself does not in terms apply itself to a breach of contract **by the Management Contractor** occasioned by the breach by a Works Contractor of a Works Contract', the Management Contractor may be unprotected against liability to the Employer, clause 3.21 being procedural only; and

— Rowe & Maw (D-38 ff) doubt whether clause 3.21 protects the Management Contractor after Final Certificate.

44. While the writer does not necessarily accept the above propositions, they cannot be lightly dismissed, and no doubt they will be asserted in litigation and arbitration in the future. Even though the suggestions are patently contrary to the policy and intention of the JCT, as expressed in MC/1 and MC/2, this might not avoid judicial decisions undermining the basis of JCT 87.

45. The suggested defect in clause 1.6.2 of Works Contract/2 could perhaps be remedied by an express obligation on the Works Contractor to admit liability to the Management Contractor in respect of the matters mentioned in clause 1.6.1 of Works Contract/2, notwithstanding that the Management Contractor is protected by clause 3.21 of JCT 87.

JCT 87: Determination of Management Contractor's Employment

46. Determination of the Management Contractor's employment under clause 7.1, 7.2 or 7.3 of JCT 87 by reason of the Management Contractor's breach or insolvency,or under clause 7.10, which enables the Employer to determine at will, **also leads to determination of the employment of all the Works Contractors.** See clause 7.9 of Works Contract/2. This could, in some circumstances, be disastrous to the Employer's interests. For example, all the Works Contractors would be able to negotiate different prices and times for completion as a condition of continuing with the Works Contract Works. Such considerations may deter removal of the Management Contractor. Therefore, the Management Contractor may be practically irremovable. Such a problem could not arise under construction management.

47. Under clause 7.4 of JCT 87, the Employer has at least a theoretical right to recover his losses from the Management Contractor in the event of determination by reason of the latter's breach or insolvency, but that may be of little benefit to the Employer in practice.

JCT 87: Management Contractor's Ambiguous Position

48. The above problems appear to be fundamental to the Management Contractor's ambiguous position under JCT 87. He is the Main Contractor, and yet is relieved from the fundamental responsibility of a Main Contractor with regard to the risk of cost over-runs and the performance of sub-contractors, being remunerated on a fee basis and protected (supposedly) by clause 3.21. Although he is paid on a fee basis, like a professional adviser, he cannot be removed at will during the Construction Period without paying him all the profit he would have made on the entire Project (clauses 7.6.5 and 7.13) and effectively removing all the Works Contractors as well. It is necessary to create a highly artificial structure, as recorded in clauses 1.7 and 3.21 of JCT 87 and clause 1.6.2 of Works Contract/2, in order to deal with the consequences of the Management Contractor's role as Main Contractor being largely nominal, while the Management Contractor still has to be obtruded between the Employer and the Works Contractors.

49. Therefore, while the Management Contractor certainly does not fulfil the role of a professional adviser, his role as Main Contractor is largely titular, as he does not assume a Main Contractor's normal risk and responsibility. In a dispute over a Works Contractor's performance, where will his interests and loyalties lie? Certainly not unambiguously with the Employer.

50. The advantages to the Employer in such a system are difficult to detect, particularly when contrasted with construction management.

Construction Management as an Alternative to Management Contracting

51. All the above serious problems with management contracting can be successfully avoided by construction management, without any discernible disadvantages.

52. In distinction to the Management Contractor, the Construction Manager should be unambiguously the Employer's professional adviser, and should have no interests adverse to the Employer. The Construction Manager may, of course be an employee of the Employer. The Employer should see to it that he can change the Construction Manager without consulting Trade Contractors. The Construction Manager's fee structure should, if possible, reward successful management, and should provide for determination of the Construction Manager's employment at the Employer's will or on short notice, without any financial commitment beyond paying fees for work done up to the time of determination. There should be no equivalent, in the Construction Manager's terms of engagement, of clauses 7.6 and 7.13 of JCT 87, giving the Construction Manager his entire projected profit on the Project if his employment is prematurely determined.

53. It is notable that there is a far wider choice of Construction Managers than Management Contractors. Management Contracts contain obligations which could only realistically be accepted by Main Contractors. A prime example is the obligation to pay Works Contractors, which is not, under the JCT Works Contract documentation, dependent on the Management Contractor being paid by the Employer. In contrast, construction management will normally involve no more than assuming a professional duty of skill, care and diligence to the Employer. Therefore, professional firms are well able to act as Construction Managers, if they have the necessary experience. The firm involved will, of course, need to ensure that its professional indemnity insurance is wide enough to cover construction management. The Employer, if he is well advised, will also be very interested in the firm's professional indemnity insurance, and should require that it shall be maintained for a number of years after completion of the Project. The same comment applies, of course, to all professional terms of engagement and collateral warranties.

54. The RICS Project Management Agreement and Conditions of Engagement could well be used as a basis for the Construction Manager's appointment.

55. The various Trade Contracts may be awarded on whatever terms and conditions of contract, including terms and conditions relating to payment, damages for delay and bonding, are considered best from the Employer's point of view. These may be JCT 80 With or Without Quantities, MF/1 in respect of mechanical and electrical Works, or others. There is no straitjacket of Works Contract Conditions imposed solely by reason of the existence of a Management Contract and the presence of a Management Contractor, and which leads in practice to interminable tripartite negotiations.

56. If a Trade Contractor is in breach, the Employer may go after him directly, without going through any Main Contractor, and may entirely control the claim. There will also be no need for any equivalent of Works Contract/3, with the resultant confusion over distinctions between design and workmanship. The Employer will be able to decide how far to delegate conduct of the claim to the Construction Manager, or anyone else.

57. It is sometimes suggested that the multiplicity of Trade Contracts directly between the Employer and Trade Contractors, and the resulting multiplicity of payments by the Employer, are disadvantages of construction management, in that they impose a more active role and more administration on the Employer. However, the Employer may delegate as much as he desires to the Construction Manager. He could even, although the writer would not recommend it, delegate to the Construction Manager power to sign Trade Contracts on the

Employer's behalf and to act as the Employer's paymaster by receiving and splitting up one periodic payment from the Employer between the Trade Contractors. Also, an Employer who does not desire an active role in his own construction Project and is not up to signing Trade Contracts and issuing a run of cheques every month upon the certification of the Construction Manager, is probably unsuitable for either management contracting or construction management. It is also to be borne in mind that there is no delay under construction management in the Trade Contractors receiving payment, which will inevitably happen if the funds pass through the hands of a Management Contractor.

58. In order to protect the Employer's position and to secure a sensible contractual structure, the Trade Contracts will need to have some construction management features in common.

59. As noted above, the JCT publishes no conditions of contract for construction management. In view of the present dominance of the JCT forms, this is in itself a serious obstacle to the use of construction management in the UK. So far, the use of construction management in this country seems to have been confined to a few large projects, such as Broadgate and Canary Wharf.

60. However, construction management is widely used in the United States, and the AIA has published a comprehensive series of construction management forms, all of which are included in the AIA Handbook. These are:

   – A1A Document A101/CM : Owner-Contractor Agreement Form : Stipulated Sum : Construction Management Edition;
   – A1A Document A201/CM : General Conditions of the Contract for Construction : Construction Management Edition;
   – A1A Document B141/CM : Standard Form of Agreement Between Owner and Architect : Construction Management Edition; and
   – A1A Document B801 : Standard Form of Agreement between Owner and Construction Manager.

61. Pending the publication of similar forms in the UK, it is submitted that acceptable construction management documentation can be achieved by relatively modest amendments to standard forms already in common use, using the AIA forms for help and guidance as necessary.

Adaptation of JCT 80 to Construction Management

62. The writer has reviewed JCT 80, with a view to amending and adding to it for

construction management purposes, and some suggestions are set out in the following paragraphs. JCT 80 is, of course, a form of contract for use with Bills of Quantities, but Bills of Quantities will often not be used in conjunction with Trade Contracts in a construction management Project. In such a case, the JCT 80 form Without Quantities could be similarly adapted.

63. Under JCT 80, all instructions, certificates etc. are given by the Architect, thus preserving a single channel of control and communication. The Construction Manager, who might, of course, also be an Architect, would be substituted in the Architect's role. The AIA CM documentation contemplates a separate role for the Architect vis-à-vis the Trade Contractors, but it is suggested that the creation of more than one channel of control and communication will usually be inappropriate. The substitution of the Construction Manager for the Architect will necessitate changes in Article 3 (Architect) of the Articles of Agreement and in clause 1.3 (Definitions).

64. In addition to the powers of the Architect to give instructions (see clause 4 (Architect's Instructions)) the Construction Manager needs to have powers with regard to scheduling, programming and co-ordination of the several Trade Contractors.

65. Clause 5.3, especially 5.3.2, will need amendment in view of the crucial need under construction management for co-ordination of the several Trade Contractors, particularly for the purpose of ensuring that their various Works Package programmes accord with one another and with the programme of the Project as a whole.

66. Just as a Main Contractor must co-ordinate his Sub-Contractors, suppliers and his own labour, the Employer, acting by his agent the Construction Manager, must co-ordinate his several Trade Contractors. If any suppliers to the Employer are involved (e.g. if the Employer is going to 'free issue' any supplies to the Trade Contractors) or any Employer's own labour (e.g. if the Employer himself is providing site services such as those to be described in the Fifth Schedule of JCT 87) the Construction Manager must also co-ordinate them. The programming and scheduling requirements are essentially the same under any form of contracting, but under construction management the Employer himself is fulfilling the role of Main Contractor. Under management contracting, the Employer bears most of the risks and responsibilities usually borne by a Main Contractor, but without having authority over the Works Contractors, who actually execute the Project.

67. As the Project will consist of several Works Packages comprised in several

separate direct Trade Contracts, the Employer may well wish to assume responsibility for Project insurance. Therefore, he may not wish to exercise his option to require the Contractor to take out clause 21.2.1 insurance against the liability of the Employer, and he may also wish to insure the Works himself under clause 22B, or clause 22C if there are existing structures, rather than requiring the Contractor to insure the Works under clause 22A. The Employer may, in fact, prefer to use a tailor-made Project insurance programme of a type suitable for major Projects, perhaps including ten year building defects insurance.

68. Customarily, of course, a Main Contractor's liability for damages for delay in completion of the Works is liquidated and ascertained by the Main Contract, as in clause 24 (Damages for non-completion). Clause 22D provides for optional insurance against the Employer's loss of liquidated damages if the time for completion of the Works is extended under clause 25.4.3 by reason of 'loss or damage occasioned by any one or more of the Specified Perils', which are defined by clause 1.3 as:

'fire, lightning, explosion, storm, tempest, flood, bursting or overflowing of water tanks, apparatus or pipes, earthquake, aircraft and other aerial devices or articles dropped therefrom, riot and civil commotion, but excluding Excepted Risks.'

The 'Excepted Risks' are defined by clause 1.3 as:

'ionizing radiations or contamination by radioactivity from any nuclear fuel or from any nuclear waste from the combustion of nuclear fuel, radioactive toxic explosive or other hazardous properties of any explosive nuclear assembly or nuclear component thereof, pressure waves caused by aircraft or other aerial devices travelling at sonic or supersonic speeds.

69. However, Sub-Contracts very often do not provide for the Sub-Contractor's liability for damages for delay in completion of the Sub-Contract Works to be liquidated and ascertained. Main Contractors generally take care to see that a delaying Sub-Contractor will be liable to indemnify the Main Contractor against **all** loss or damage falling on the Main Contractor as a result of the Sub-Contractor's delay. This will include any liquidated and ascertained damages payable under the Main Contract for delay in completion of the Main Works, but will not necessarily be limited to such damages. The Main Contractor will frequently himself suffer disruption and expense because of the Sub-Contractor's delay. This may include disruption claims from other Sub-Contractors, resulting from the delay of the Sub-Contractor in default.

70. Under construction management, those who might have been Sub-Contractors to a single Main Contractor under the usual contractual pyramid structure adopted in lump sum, 'design and build' or management contracting, will each have direct Trade Contracts with the Employer. There is no logical reason why the Employer should lose out as a result. Therefore, the Employer should consider imposing on the Trade Contractors damages at large, or liquidated and ascertained damages by reference to the contract value of the entire Project, rather than the contract value of the particular Trade Contractor's Works.

71. It will frequently happen that some Trade Contractors will need to complete their Works by contractual completion dates long before the intended completion date of the Project. In that case, the Employer's loss by reason of a relevant Trade Contractor's delay may be of two different kinds. In the period before the intended completion date of the Project, his loss will be likely to be the disruption claims asserted by the other Trade Contractors. In the period after that date, his loss will be likely to be loss of the income of the Project, e.g. rent from a commercial building, or production from a process or manufacturing plant. As the former loss may not be realistically estimable in advance, but the latter often is, it might be logical to provide for damages at large during the period up to the intended completion date of the Project, and thereafter for liquidated and ascertained damages.

72. Under a 'back-to-back' Sub-Contract, a wise Main Contractor will pass on to the Sub-Contractor unlimited liability for disrupting the Main Contractor, the Main Works and other Sub-Contractors. Taking that example, the Employer under construction management should impose equivalent liabilities upon his several Trade Contractors in relation to the entire Project, not only their own Works. This could be achieved by a provision in the Trade Contracts to the effect that the Trade Contractor shall indemnify the Employer against all loss or damage flowing from his breach, including particularly loss or damage arising from disruption of other Trade Contracts relating to the Project. Such disruption might not only occur by reason of late completion of the Trade Contractor's Works, but by failure to meet intermediate dates in the relevant programme.

73. Clearly, no single Trade Contractor will be given possession of the Site. All the Trade Contractors will have to share it. Each must be given appropriate access and opportunity to execute his Works, and each must allow such access and opportunity to the other Trade Contractors. Clause 29 will have to reflect this. Site services (cf, JCT 87, Fifth Schedule) will have to be dealt with in some way, either by one of the Trade Contractors providing them, or by the Construction

Manager or the Employer providing them.

74. It may be valuable for the Construction Manager, on behalf of the Employer, to be able to require a Contractor to accelerate his Works. Clause 3.6 of JCT 87 could serve as a precedent for a suitable acceleration clause, suitably amended and elaborated so that it does not present the Employer with only a 'take it or leave it' choice. Acceleration powers in construction management may be particularly appropriate, bearing in mind the great importance of the Construction Manager's co-ordinating and scheduling role.

75. Non-co-operation with the Construction Manager or other Trade Contractors might be added as a ground for determination under clause 27.1, as such co-operation is crucial to construction management.

76. Normally, the Employer will also require the Defects Liability Period for the whole Project to be uniform, necessitating changes to clause 17. The intermediate period between completion of the relevant Trade Contractor's Works and completion of the Project will have to be covered in respect of defects liability, loss or damage, etc.

77. Consolidation of arbitrations with Trade Contractors under clause 41 would also be desirable.

78. Although ignored by all the JCT forms of contract, bonding in respect of the Main Contractor's performance of the Contract by a bank or surety company in favour of the Employer is, of course, a common and important feature of construction contracts. Such bonding is customarily for 10% of the Contract Sum, often reducing to 5% on Practical Completion, with final release on settlement under the Final Certificate or its equivalent. Under management contracting, the Management Contractor's performance bond will, naturally, only answer for the Management Contractor's liability under the Contract. Bonding does not affect the 'pay-when-paid' nature of the Management Contractor's liability under clause 3.21 in respect of Works Contractors' breaches. The Employer will usually also have entered into Works Contract/3 with the Works Contractor. However, if the Works Contractor becomes insolvent and the Employer has no bonding from the Works Contractor in respect of the Works Contractor's performance of Works Contract/3, the Employer may, because of clause 3.21 of JCT 87, have no effective remedy against the Management Contractor or the Management Contractor's performance bondsman. Therefore, the Employer may wish to consider direct bonding from the Works Contractor in respect of the Works Contractor's performance of Works Contract/3, but that may lead to double bonding costs.

79.   Under construction management, all performance bonding by Trade Contractors will be direct to the Employer, who may wish to consider a higher percentage in each case than 10%, bearing in mind that each Trade Contractor may delay and disrupt all the rest. However, there will be no need for any double bonding costs. As in any other form of contracting, the Employer should, where appropriate, obtain parent company guarantees.

80.   The JCT forms of contract also ignore the question of collateral warranties, with the exception of collateral warranties, such as Works Contract/3, by sub-contractors to Employers.

81.   Collateral warranties by a Main Contractors to a Fund usually provide for the Fund to take over as the Main Contractor's Employer in certain eventualities, such as termination by the Fund of the Employer's financing arrangements, or impending determination by the Main Contractor of his employment under the Main Contract by reason of the Employer's breach. In the latter case, the collateral warranty will normally require the Main Contractor to give to the Fund a warning notice of the impending determination.

82.   There is no reason why the several Trade Contractors engaged in a construction management Project should not give the like collateral warranties to the Fund. However, the Fund will need to avoid taking over some but not all of the Trade Contracts as Employer, which would create an impossible tangle. Therefore, the Trade Contractors' collateral warranties to the Fund could provide that, if the Fund becomes entitled to take over as Employer in respect of any one of the Trade Contracts, the Fund shall be entitled to take over all the Trade Contracts as Employer.

The *Broadgate* and *Zimmcor* Cases

83.   In a paper on construction management, it is perhaps as well to mention the cases of *Rosehaugh Stanhope (Broadgate Phase 6) PLC v Redpath Dorman Long Limited* (1990) 50 BLR 69, CA and *Beaufort House Development Limited v Zimmcor (International) Inc* (1990) 50 BLR 91. These cases turned solely on the terms of the particular Contracts and the writer does not consider that they establish anything particularly relevant to construction management. They are equally applicable to all forms of contracting, as pointed out in an article entitled 'An Attack on Construction Management?' by Henry Sherman and Nigel Proctor of McKenna & Co in the Property Law Journal, Vol. XV, No. 5, p. 8, October 1990.

Conclusion

84. In a Project which is suitable either for construction management or management contracting, there seem to be patent advantages to the Employer in choosing construction management. The disadvantage of choosing construction management appears to be that there are no suitable published forms of contract. However, the present published forms can be adapted without much difficulty. The published forms are almost invariably subjected to significant amendment in any event, other than in the case of small Projects which would not be suitable for management contracting or construction management anyway.

Postcript

85. Since writing this paper, the writer has had the great advantage of reading 'Construction Management Forum: Report and Guidance', published by the Centre for Strategic Studies in Construction of the University of Reading in January 1991. That extremely useful document naturally approaches the subject of construction management on a much broader front and with less legal detail than this paper, but comes to similar conclusions.

86. Again since writing this paper, the writer has received from the University of Southampton a copy of a Science and Engineering Research Council research grant report dated December 1989 and entitled 'Roles, Responsibilities and Risks in Management Contracts', prepared by staff of the University, which contains a wealth of very useful and relevant research information and comment.

FORMS

FORM 1: ARCHITECT'S APPOINTMENT, INCORPORATING THE RIBA ARCHITECT'S APPOINTMENT, 1982 EDITION

DATED _____ **199**

[   ]

('the Client')

-and-

[   ]

('the Architect')

ARCHITECT'S APPOINTMENT

for professional services
relating to the construction of

[   ]

✷ An asterisk indicates that there is a relevant note in the commentary upon this Form.

THIS AGREEMENT made the   day of   One thousand nine
hundred and ninety-

BETWEEN:

(1)

[of] **OR** [whose registered office is at]

('the Client'); and

(2)

[of] **OR** [whose registered office is at]

('the Architect');

INCORPORATES the Royal Institute of British Architects' Architect's Appointment (1982
Edition, February 1990 Revision) as amended in July 1990, and as amended by the Schedule
of Services and Fees. 'The Works' are the construction of

at

IN WITNESS whereof the parties have executed this Deed in duplicate on the date first stated
above.

SCHEDULE OF SERVICES AND FEES✱

S2   – SPECIAL CONDITIONS

1.1✱   The Architect shall maintain professional indemnity insurance covering (inter alia) all
liability hereunder upon customary and usual terms and conditions prevailing for the
time being in the insurance market, and with reputable insurers lawfully carrying on
such insurance business in the United Kingdom, in an amount of not less than
                    pounds (£   ) for any one occurrence or series of
occurrences arising out of any one event for a period beginning now and ending 15
(fifteen) years after the date of practical completion of the Works for the purposes of
the building contract, provided always that such insurance is available at commercially
reasonable rates. The said terms and conditions shall not include any term or condition
to the effect that the Architect must discharge any liability before being entitled to

recover from the insurers, or any other term or condition which might adversely affect the rights of any person to recover from the insurers pursuant to the Third Parties (Rights Against Insurers) Act 1930, or any amendment or re-enactment thereof. The Architect shall not, without the prior approval in writing of the Client, settle or compromise with the insurers any claim which the Architect may have against the insurers and which relates to a claim by the Client against the Architect, or by any act or omission lose or prejudice the Architect's right to make or proceed with such a claim against the insurers.

1.2   Any increased or additional premium required by insurers by reason of the Architect's own claims record or other acts, omissions, matters or things particular to the Architect shall be deemed to be within commercially reasonable rates.

1.3   The Architect shall immediately inform the Client if such insurance ceases to be available at commercially reasonable rates in order that the Architect and the Client can discuss means of best protecting the respective positions of the Client and the Architect in respect of the Works in the absence of such insurance.

1.4   The Architect shall fully co-operate with any measures reasonably required by the Client, including (without limitation) completing any proposals for insurance and associated documents, maintaining such insurance at rates above commercially reasonable rates if the Client undertakes in writing to reimburse the Architect in respect of the net cost of such insurance to the Architect above commercially reasonable rates or, if the Client effects such insurance at rates at or above commercially reasonable rates, reimbursing the Client in respect of what the net cost of such insurance to the Client would have been at commercially reasonable rates.

1.5   As and when reasonably requested to do so by the Client, the Architect shall produce for inspection documentary evidence (including, if required by the Client, the original of the relevant insurance documents) that his professional indemnity insurance is being maintained.

1.6   The above obligations in respect of professional indemnity insurance shall continue notwithstanding termination of this Agreement for any reason whatsoever, including (without limitation) breach by the Client.

2.1✱   The Architect shall within 7 (seven) working days of the Client's request so to do:

2.1.1   execute, in favour of any persons who have entered or shall enter into an agreement for the provision of finance in connection with the Works, a Deed in the form annexed as Annexe 'A', or a similar form reasonably required by the Client, and deliver the same to the Client, together with a guarantee in the form annexed as Annexe 'C', or a similar

form reasonably required by the Client, from [   ] and [   ]
OR [the ultimate parent company of the Architect, namely [   ]
[Limited] **OR** [PLC]], in respect of the Architect's obligations pursuant to such Deed;
and

2.1.2   execute, in favour of any persons who have acquired or shall acquire any interest in
or over the Works or any part thereof, a Deed in the form annexed as Annexe 'B', or
a similar form reasonably required by the Client, and deliver the same to the Client,
together with a guarantee in the form annexed as Annexe 'C', or a similar form
reasonably required by the Client, from [   ] and [   ] **OR**
[the ultimate parent company of the Architect, namely [   ] [Limited]
OR [PLC]], in respect of the Architect's obligations pursuant to such Deed.

2.2   The above obligation for the provision of Deeds and guarantees in favour of third
parties shall continue notwithstanding termination of this Agreement for any reason
whatsoever, including (without limitation) breach by the Client. However, any such
Deed given after such termination shall be amended by the Client so as to refer to the
fact and date of such termination, to omit any obligation to continue to exercise skill,
care and diligence after such termination, and to omit any provision enabling a third
party to assume the position of the Client.

3.＊   Upon execution of this Deed, the Architect shall deliver to the Client a guarantee from
[   ] and [   ] **OR** parent company guarantee from its
ultimate parent company, namely [   ] [Limited] **OR** [PLC], for its
performance of this Deed in the form annexed as Annexe 'D'.

4.   Clause 3.1 shall be deleted, and the following substituted:

'3.1.1   The Architect has exercised, and will exercise, reasonable skill, care and diligence in
conformity with the normal standards of the Architect's profession.

3.1.2   Without prejudice to the generality of clause 3.1.1, the Architect warrants:

3.1.2.1   that he has not specified and will not specify for use;

3.1.2.2   that he has exercised and will continue to exercise reasonable skill, care and
diligence to see that there are not used;

3.1.2.3   that he is not aware and has no reason to suspect or believe that there have
been or will be used; and

3.1.2.4   that he will promptly notify the Client in writing if he becomes aware or has reason to suspect or believe that there have been or will be used;

in or in connection with the Works, any of the materials or substances identified in clause 3.1.3.

3.1.3   The said materials or substances are:

3.1.3.1   high alumina cement in structural elements;

3.1.3.2   wood wool slabs in permanent formwork to concrete;

3.1.3.3   calcium chloride in admixtures for use in reinforced concrete;

3.1.3.4   asbestos products;

3.1.3.5   naturally occurring aggregates for use in reinforced concrete which do not comply with British Standard 882: 1983 and/or naturally occurring aggregates for use in concrete which do not comply with British Standard 8110: 1985.'

5.   Clause 3.5 shall be deleted, and the following substituted:

'3.5.1   Consultants may be nominated by either the Client or the Architect, subject to acceptance by each party, such acceptance not to be unreasonably withheld or delayed.

3.5.2   Where the Client accepts the nomination of a consultant to be employed by the Architect, the Client's prior approval in writing shall also be required in respect of the terms and conditions of such consultant's engagement by the Architect, which shall be by Deed.

3.5.3   The Architect shall be solely responsible for the remuneration of such consultants employed by him, and shall not amend the terms and conditions of any such engagements, or depart therefrom, without the Client's prior approval in writing.

3.5.4   The Architect shall procure that each such consultant employed by the Architect, upon being engaged by the Architect, and as a condition precedent to the relevant engagement, shall acknowledge by Deed to the Client (in form and substance approved or reasonably required by the Client) that such consultant owes a duty direct to the Client to have exercised, and to exercise, reasonable skill, care and diligence in conformity with the normal standards of such consultant's profession in carrying out the relevant engagement.'

6.   Clause 3.6 shall be deleted, and the following substituted:

'Where the Client employs consultants directly, the Client will hold each consultant, and not the Architect, responsible for the competence, general inspection and performance of the work entrusted to that consultant; where the Architect employs consultants, the Client shall be entitled to hold the Architect, as well as each consultant, responsible for the competence, general inspection and performance of the work entrusted to that consultant; provided that in relation to the execution of such work under the contract between the Client and the contractor nothing in this clause shall affect any responsibility of the Architect for issuing instructions or for other functions ascribed to the Architect under that contract.'

7.   Add to clause 3.15:

'The Architect shall promptly supply the Client with conveniently reproducible copies of all such documents and drawings, as and when they are intended to be issued or used.'

8.   Clause 3.16 shall be deleted, and the following substituted:

'The Client will be entitled to reproduce all such documents and drawings and the Architect's design by proceeding to execute the project, provided that any fees due to the Architect have been paid or tendered. This entitlement will also apply to the maintenance, repair , renewal, letting, promotion, advertisement and/or extension of the Works.'

9.   Add to clause 3.19:

'such consent not to be unreasonably withheld or delayed.'

10.   Clause 3.23 shall be deleted, and the following substituted:

'The Architect's appointment may be terminated by the Client on the expiry of reasonable notice given in writing.'

11.   The proviso to clause 3.26 shall be deleted.

12.   In clause 4.23B, the words 'the architect's principal bank' shall be deleted and '[   ] Bank PLC' substituted.

S3 — CONDITIONS NOT TO APPLY OR TO APPLY AS AMENDED:

3.1
3.5
3.6
3.15
3.16
3.17
3.18
3.19
3.23
3.26
3.26S
3.28S
3.28NI
4.23B

COMMENTARY

Schedule of Services and Fees

The Schedule should be duly completed as published by the RIBA, subject to the incorporation of Sections S2 and S3 above.

Section S2
Special Condition 1: Professional Indemnity Insurance

See commentary upon clause 9 of Form 6.

Special Condition 2: Collateral Warranties

For Annexes 'A', 'B' and 'C', see Forms 6, 7, 8 and 9.

Special Condition 3: Personal Guarantees or Parent Company Guarantee

For Annexe 'D', see Forms 10 and 11.

FORM 1A: ALTERNATIVE ARCHITECT'S APPOINTMENT, INCORPORATING THE RIBA STANDARD FORM OF AGREEMENT FOR THE APPOINTMENT OF AN ARCHITECT (SFA/92)

DATED _____ **199**

[   ]

('the Client')

-and-

[   ]

('the Architect')

ARCHITECT'S APPOINTMENT

for professional services
relating to the construction of
[   ]

✳ An asterisk indicates that there is a relevant note in the commentary upon this Form.

THIS AGREEMENT made the   day of   One thousand nine hundred and ninety-

BETWEEN:

(1)

[of] **OR** [whose registered office is at]

('the Client'); and

(2)

[of] **OR** [whose registered office is at]

('the Architect');

INCORPORATES the parts of the Standard Form of Agreement for the Appointment of an Architect (SFA/92), RIBA Edition, as printed in July 1992, indicated in the Appendix, as thereby amended.

IN WITNESS whereof the parties have executed this Deed in duplicate on the date first stated above.

APPENDIX

Incorporation of SFA/92

The parts of SFA/92 incorporated into this Agreement are as follows:
- Memorandum of Agreement (Alternative version for execution as a Deed under the law of England and Wales), as hereby amended;
- Definitions;
- Schedule One (Information to be supplied by Client), as annexed;✹
- Schedule Two (Services to be Provided by Architect), as annexed;✹
- Conditions of Appointment, as hereby amended;
- Schedule Three (Fees and Expenses), as annexed; and✹
- Schedule Four (Appointment of Consultants, Specialists and Site Staff), as annexed.✹

Amendments to Memorandum of Agreement

Recital A

The Project is the construction of

Recital B

The Site is

Clauses 5 and 6

Clauses 5 and 6 shall be deleted.

Amendments to Conditions of Appointment

Condition 1.1: Governing law/interpretation

In condition 1.1.1, '[Northern Ireland] [Scotland]' shall be deleted.

Condition 1.2.1: Duty of Care

Condition 1.2.1 shall be deleted, and the following substituted:

'1.2.1.1 The Architect has exercised, and will exercise, reasonable skill, care and diligence in conformity with the normal standards of the Architect's profession.

1.2.1.2 Without prejudice to the generality of condition 1.2.1.1, the Architect warrants:

    1.2.1.2.1   that he has not specified and will not specify for use;

    1.2.1.2.2   that he has exercised and will continue to exercise reasonable skill, care and diligence to see that there are not used;

    1.2.1.2.3   that he is not aware and has no reason to suspect or believe that there have been or will be used; and

1.2.1.2.4   that he will promptly notify the Client in writing if he becomes aware or has reason to suspect or believe that there have been or will be used;

in or in connection with the Works, any of the materials or substances identified in condition 1.2.1.3.

1.2.1.3   The said materials or substances are:

     1.2.1.3.1   high alumina cement in structural elements;

     1.2.1.3.2   wood wool slabs in permanent formwork to concrete;

     1.2.1.3.3   calcium chloride in admixtures for use in reinforced concrete;

     1.2.1.3.4   asbestos products;

     1.2.1.3.5   naturally occurring aggregates for use in reinforced concrete which do not comply with British Standard 882: 1983 and/or naturally occurring aggregates for use in concrete which do not comply with British Standard 8110: 1985.'

Condition 1.4.1: Assignment

Add to condition 1.4.1:

'such consent not to be unreasonably withheld or delayed.'

Condition 1.5.5: Fee variation

Condition 1.5.5 shall be deleted, and the following substituted:

'1.5.5   Where any change is made to the Architect's Services, the Procurement Method, the Client's Requirements, the Budget, or the Timetable, the fees specified in Schedule Three shall be varied.'

Condition 1.5.15: No setoff

Condition 1.5.15 shall be deleted.

Condition 1.6.5: Termination

Condition 1.6.5 shall be deleted, and the following substituted:

'1.6.5   The Appointment may be terminated by the Client on the expiry of reasonable notice in writing.'

Condition 1.7.1: Copyright

Add to condition 1.7.1:

'The Architect shall promptly supply the Client with conveniently reproducible copies of all such documents and drawings, as and when they are intended to be issued or used.'

Condition 1.8.1: Arbitration

The proviso to condition 1.8.1 shall be deleted.

Condition 1.8.1S shall be deleted.

Condition 2.2.2

Condition 2.2.2 shall be deleted.

Condition 2.2.3: Collateral Agreements

Condition 2.2.3 shall be deleted.

Condition 2.3: Copyright

Condition 2.3.1 shall be deleted, and the following substituted:

'2.3.1   Notwithstanding the provisions of condition 1.7.1, the Client shall be entitled to reproduce all documents and drawings referred to therein and the Architect's design by proceeding to execute the Project, provided that any fees, expenses and disbursements due to the Architect have been paid or tendered. This entitlement shall also apply to the maintenance, repair, renewal, letting, promotion, advertisement and/or extension of the Works.'

Conditions 2.3.2, 2.3.3 and 2.3.4 shall be deleted.

Condition 3.2.4: Collateral Agreements

Condition 3.2.4 shall be deleted.

Condition 3.2.5: Instructions

Condition 3.2.5 shall be deleted.

Condition 4.1.1: Nomination

Add to condition 4.1.1:

'such acceptance not to be unreasonably withheld or delayed.'

Condition 4.1.4: Collateral Agreements

Condition 4.1.4 shall be deleted.

Condition 4.1.5: Lead Consultant

Condition 4.1.5 shall be deleted, and the following substituted:

'4.1.5   [The Client shall appoint and give authority to the Architect as Lead Consultant in relation to all consultants however employed.] The Architect shall coordinate and integrate into the overall design the services of the consultants [and require reports from the consultants].'

Condition 4.1.6

Condition 4.1.6 shall be deleted, and the following substituted:

'4.1.6   The Client shall procure that the provisions of condition 4.1.5 above are incorporated into the conditions of appointment of all consultants however employed and shall provide a copy of such conditions of appointment to the Architect, less such details of fees and other matters which the Client considers should be held confidential between the Client and such consultants.'

Condition 4.2.1: Nomination

Condition 4.2.1 shall be deleted, and the following substituted:

'4.2.1   A Specialist who is to be employed directly by the Client or indirectly through the contractor to design any part of the Works may be nominated by either the Architect, subject to acceptance by the Client, or by the Client.'

Condition 4.2.3: Collateral Agreements

Condition 4.2.3 shall be deleted.

Additional Conditions

The following conditions shall be added:

'5.1:   Professional Indemnity Insurance*

5.1.1   The Architect shall maintain professional indemnity insurance covering (inter alia) all liability hereunder upon customary and usual terms and conditions prevailing for the time being in the insurance market, and with reputable insurers lawfully carrying on such insurance business in the United Kingdom, in an amount of not less than   pounds (£   ) for any one occurrence or series of occurrences arising out of any one event for a period beginning now and ending 15 (fifteen) years after the date of practical completion of the Works for the purposes of the building contract, provided always that such insurance is available at commercially reasonable rates. The said terms and conditions shall not include any term or condition to the effect that the Architect must discharge any liability before being entitled to recover from the insurers, or any other term or condition which might adversely affect the rights of any person to recover from the insurers pursuant to the Third Parties (Rights Against Insurers) Act 1930, or any amendment or re-enactment thereof. The Architect shall not, without the prior approval in writing of the Client, settle or compromise with the insurers any claim which the Architect may have against the insurers and which relates to a claim by the Client against the Architect, or by any act or omission lose or prejudice the Architect's right to make or proceed with such a claim against the insurers.

5.1.2 Any increased or additional premium required by insurers by reason of the Architect's own claims record or other acts, omissions, matters or things particular to the Architect shall be deemed to be within commercially reasonable rates.

5.1.3 The Architect shall immediately inform the Client if such insurance ceases to be available at commercially reasonable rates in order that the Architect and the Client can discuss means of best protecting the respective positions of the Client and the Architect in respect of the Works in the absence of such insurance.

5.1.4 The Architect shall fully co-operate with any measures reasonably required by the Client, including (without limitation) completing any proposals for insurance and associated documents, maintaining such insurance at rates above commercially reasonable rates if the Client undertakes in writing to reimburse the Architect in respect of the net cost of such insurance to the Architect above commercially reasonable rates or, if the Client effects such insurance at rates at or above commercially reasonable rates, reimbursing the Client in respect of what the net cost of such insurance to the Client would have been at commercially reasonable rates.

5.1.5 As and when reasonably requested to do so by the Client, the Architect shall produce for inspection documentary evidence (including, if required by the Client, the original of the relevant insurance documents) that his professional indemnity insurance is being maintained.

5.1.6 The above obligations in respect of professional indemnity insurance shall continue notwithstanding termination of this Agreement for any reason whatsoever, including (without limitation) breach by the Client.

5.2: Collateral Warranties*

5.2.1 The Architect shall within 7 (seven) working days of the Client's request so to do:-

5.2.1.1 execute, in favour of any persons who have entered or shall enter into an agreement for the provision of finance in connection with the Works, a Deed in the form annexed as Annexe 'A', or a similar form reasonably required by the Client, and deliver the same to the Client, together with a guarantee in the form annexed as Annexe 'C', or a similar form reasonably required by the Client, from [   ] and

[   ] **OR** [the ultimate parent company of the Architect, namely
[   ] [Limited] **OR** [PLC]], in respect of the Architect's obligations
pursuant to such Deed; and

5.2.1.2   execute, in favour of any persons who have acquired or shall acquire any interest in
or over the Works or any part thereof, a Deed in the form annexed as Annexe 'B', or
a similar form reasonably required by the Client, and deliver the same to the Client,
together with a guarantee in the form annexed as Annexe 'C', or a similar form
reasonably required by the Client, from [   ] and
[   ] **OR** [the ultimate parent company of the Architect, namely
[   ] [Limited] **OR** [PLC]], in respect of the Architect's obligations
pursuant to such Deed.

5.2.2   The above obligation for the provision of Deeds and guarantees in favour of third
parties shall continue notwithstanding termination of this Agreement for any reason
whatsoever, including (without limitation) breach by the Client. However, any such
Deed given after such termination shall be amended by the Client so as to refer to the
fact and date of such termination, to omit any obligation to continue to exercise skill,
care and diligence after such termination, and to omit any provision enabling a third
party to assume the position of the Client.

5.3:   [Personal Guarantees] OR [Parent Company Guarantee]✳

5.3.1   Upon execution of this Deed, the Architect shall deliver to the Client a guarantee from
[   ] and [   ] **OR** parent company guarantee
from its ultimate parent company, namely [   ] [Limited] **OR**
[PLC], for its performance of this Deed in the form annexed as Annexe 'D'.'

COMMENTARY

Schedules

See paragraph 13 of Chapter 1.

The relevant Schedules, as published by the RIBA, should be duly completed. If necessary, they should be
amended and expanded to reflect the actual situation and the services to be provided.

Additional Conditions

Condition 5.1: Professional Indemnity Insurance

See commentary upon clause 9 of Form 6.

Condition 5.2: Collateral Warranties

For Annexes 'A', 'B' and 'C', see Forms 6, 7, 8 and 9.

Condition 5.3: Personal Guarantees or Parent Company Guarantee

For Annexe 'D', see Forms 10 and 11.

FORM 2: QUANTITY SURVEYOR'S APPOINTMENT, INCORPORATING THE RICS QUANTITY SURVEYOR'S APPOINTMENT

DATED _____ **199**

[   ]

('the Employer')

-and-

[   ]

('the Quantity Surveyor')

QUANTITY SURVEYOR'S APPOINTMENT

for professional services
relating to the construction of

[   ]

✱ An asterisk indicates that there is a relevant note in the commentary upon this Form.

THIS AGREEMENT made the   day of   One thousand nine hundred and ninety-

BETWEEN:

(1)

[of] **OR** [whose registered office is at]

('the Employer'); and

(2)

[of] **OR** [whose registered office is at]

('the Quantity Surveyor');

INCORPORATES the February 1992 Form of Agreement and Terms and Conditions for the Appointment of a Quantity Surveyor issued by the Royal Institution of Chartered Surveyors, as amended by the Schedule.

IN WITNESS whereof the parties have executed this Deed in duplicate on the date first stated above.

SCHEDULE

Form of Agreement

Paragraph (iii) The Works

'The Works' are the construction of

at

Paragraph (iv) The Fee

The Fee is as follows:

Paragraph (v) Payment of the Fee

The Fee is payable by Instalments as follows:

Paragraph (vi) Professional Indemnity Insurance✱

Paragraph (vi) shall be deleted.

Terms and Conditions

Clause 1.0 Payment for the Quantity Surveyor's Services
Clause 1.2

Clause 1.2 shall be deleted.

Clause 1.4

The rate of interest to be inserted is   % (per cent).

Clause 1.5

Payment for the additional work mentioned in clause 1.5 shall be upon the following basis:

Clause 1.9

Clause 1.9 shall be deleted.

Clause 3.0 Insurance✱

Clause 3.0 shall be deleted.

Clause 6.0 Suspension or Termination by the Quantity Surveyor
Clause 6.3

Delete 'or threatened'.

Delete 'or any judgment against the Employer remains unsatisfied for more than 14 days'.

Clause 8.0 Copyright

Clause 8.0 shall be deleted, and the following substituted:

'Unless otherwise agreed in writing, the Quantity Surveyor shall retain copyright in all documents prepared by the Quantity Surveyor, and ownership thereof except in respect of copies supplied to the Employer. The Quantity Surveyor shall promptly supply the Employer with conveniently reproducible copies of all such documents, as and when they are intended to be issued or used. The Employer shall be entitled to reproduce the same for the purposes of the execution, maintenance, repair, renewal, letting, promotion, advertisement and/or extension of the Works.'

Clause 9.0 Liability of the Quantity Surveyor

Clause 9.0 shall be deleted, and the following substituted:

'9.1   The Quantity Surveyor has exercised, and will exercise, reasonable skill, care and diligence in conformity with the normal standards of the Quantity Surveyor's profession.

9.2   Without prejudice to the generality of clause 9.1, the Quantity Surveyor warrants:

    9.2.1   that he has not specified and will not specify for use;

    9.2.2   that he has exercised and will continue to exercise reasonable skill, care and diligence to see that there are not used;

    9.2.3   that he is not aware and has no reason to suspect or believe that there have been or will be used; and

    9.2.4   that he will promptly notify the Employer in writing if he becomes aware or has reason to suspect or believe that there have been or will be used;

in or in connection with the Works, any of the materials or substances identified in clause 9.3.

9.3   The said materials or substances are:

9.3.1   high alumina cement in structural elements;

9.3.2   wood wool slabs in permanent formwork to concrete;

9.3.3   calcium chloride in admixtures for use in reinforced concrete;

9.3.4   asbestos products;

9.3.5   naturally occurring aggregates for use in reinforced concrete which do not comply with British Standard 882: 1983 and/or naturally occurring aggregates for use in concrete which do not comply with British Standard 8110: 1985.'

The following clauses shall be added:

'13.0   Professional Indemnity Insurance✳

13.1   The Quantity Surveyor shall maintain professional indemnity insurance covering (inter alia) all liability hereunder upon customary and usual terms and conditions prevailing for the time being in the insurance market, and with reputable insurers lawfully carrying on such insurance business in the United Kingdom, in an amount of not less than   pounds (£   ) for any one occurrence or series of occurrences arising out of any one event for a period beginning now and ending 15 (fifteen) years after the date of practical completion of the Works for the purposes of the building contract, provided always that such insurance is available at commercially reasonable rates. The said terms and conditions shall not include any term or condition to the effect that the Quantity Surveyor must discharge any liability before being entitled to recover from the insurers, or any other term or condition which might adversely affect the rights of any person to recover from the insurers pursuant to the Third Parties (Rights Against Insurers) Act 1930, or any amendment or re-enactment thereof. The Quantity Surveyor shall not, without the prior approval in writing of the Employer, settle or compromise with the insurers any claim which the Quantity Surveyor may have against the insurers and which relates to a claim by the Employer against the Quantity Surveyor, or by any act or omission lose or prejudice the Quantity Surveyor's right to make or proceed with such a claim against the insurers.

13.2   Any increased or additional premium required by insurers by reason of the Quantity Surveyor's own claims record or other acts, omissions, matters or things particular to the Quantity Surveyor shall be deemed to be within commercially reasonable rates.

13.3   The Quantity Surveyor shall immediately inform the Employer if such insurance ceases to be available at commercially reasonable rates in order that the Quantity Surveyor and the Employer can discuss means of best protecting the respective positions of the Employer and the Quantity Surveyor in respect of the Works in the absence of such insurance.

13.4   The Quantity Surveyor shall fully co-operate with any measures reasonably required by the Employer, including (without limitation) completing any proposals for insurance and associated documents, maintaining such insurance at rates above commercially reasonable rates if the Employer undertakes in writing to reimburse the Quantity Surveyor in respect of the net cost of such insurance to the Quantity Surveyor above commercially reasonable rates or, if the Employer effects such insurance at rates at or above commercially reasonable rates, reimbursing the Employer in respect of what the net cost of such insurance to the Employer would have been at commercially reasonable rates.

13.5   As and when reasonably requested to do so by the Employer, the Quantity Surveyor shall produce for inspection documentary evidence (including, if required by the Employer, the original of the relevant insurance documents) that his professional indemnity insurance is being maintained.

13.6   The above obligations in respect of professional indemnity insurance shall continue notwithstanding termination of this Agreement for any reason whatsoever, including (without limitation) breach by the Employer.

14.0   Collateral Warranties✱

14.1   The Quantity Surveyor shall within 7 (seven) working days of the Employer's request so to do:

14.1.1   execute, in favour of any persons who have entered or shall enter into an agreement for the provision of finance in connection with the Works, a Deed in the form annexed as Annexe 'A', or a similar form reasonably required by the Employer, and deliver the same to the Employer, together with a guarantee in the form annexed as Annexe 'C', or a similar form reasonably required by the Employer, from [   ] and [   ] **OR** [the ultimate parent company of the Quantity Surveyor, namely [   ] [Limited] **OR** [PLC]], in respect of the Quantity Surveyor's obligations pursuant to such Deed; and

14.1.2   execute, in favour of any persons who have acquired or shall acquire any interest in or over the Works or any part thereof, a Deed in the form annexed as Annexe 'B', or a similar form reasonably required by the Employer, and deliver the same to the Employer, together with a guarantee in the form annexed as Annexe 'C', or a similar form reasonably required by the Employer, from [   ] and [   ] **OR** [the ultimate parent company of the Quantity Surveyor, namely [   ] [Limited] **OR** [PLC]], in respect of the Quantity Surveyor's obligations pursuant to such Deed.

14.2   The above obligation for the provision of Deeds and guarantees in favour of third parties shall continue notwithstanding termination of this Agreement for any reason whatsoever, including (without limitation) breach by the Employer. However, any such Deed given after such termination shall be amended by the Employer so as to refer to the fact and date of such termination, to omit any obligation to continue to exercise skill, care and diligence after such termination, and to omit any provision enabling a third party to assume the position of the Employer.

15.0   [Personal Guarantees] OR [Parent Company Guarantee]✴

Upon execution of this Deed, the Quantity Surveyor shall deliver to the Employer a guarantee from [   ] and [   ] **OR** parent company guarantee from its ultimate parent company, namely [   ] [Limited] **OR** [PLC], for its performance of this Deed in the form annexed as Annexe 'D'.'

COMMENTARY

Form of Agreement
Paragraph (vi) Professional Indemnity Insurance

See clause 13 of the Terms and Conditions

Terms and Conditions
Clause 3.0 Insurance

See clause 13 of the Terms and Conditions.

Clause 13: Professional Indemnity Insurance

See commentary upon clause 9 of Form 6.

Clause 14: Collateral Warranties

For Annexes 'A', 'B' and 'C', see Forms 6, 7, 8 and 9.

Clause 15: Personal Guarantees or Parent Company Guarantee

For Annexe 'D', see Forms 10 and 11.

FORM 3: STRUCTURAL ENGINEER'S APPOINTMENT, INCORPORATING ACE AGREEMENT 3 (1984)

DATED _____ **199**

[   ]

('the Client')

-and-

[   ]

('the Consulting Engineer')

STRUCTURAL ENGINEER'S APPOINTMENT

for professional services
relating to the construction of
[   ]

✻ An asterisk indicates that there is a relevant note in the commentary upon this Form.

THIS AGREEMENT made the   day of   One thousand nine hundred and ninety-

BETWEEN:

(1)

[of] **OR** [whose registered office is at]

('the Client'); and

(2)

[of] **OR** [whose registered office is at]

('the Consulting Engineer');

INCORPORATES the recitals and Articles 1, 2, 3, 4 and 5 of the Memorandum of Agreement and the Conditions of Engagement included in The Association of Consulting Engineers' 'Conditions of Engagement – Agreement 3(1984) – For Structural Engineering Work where an Architect is Appointed by the Client – Harmonized with the "Architect's Appointment"', as printed in 1990, and as amended in August 1990 and by the Schedule.

IN WITNESS whereof the parties have executed this Deed in duplicate on the date first stated above.

SCHEDULE

Recitals to the Memorandum of Agreement

The name and address of the Architect are

The Project is the construction of

'The Works' are

Memorandum of Agreement✱

Articles 3, 4 and 5 of the Memorandum of Agreement shall be completed as follows:

Amendments to Conditions of Engagement

1.	Clause 2.3 shall be deleted, and the following substituted:

	'If at any time the Client so decides, he may by notice in writing to the Consulting Engineer terminate the Consulting Engineer's appointment under this Agreement, provided that the Client may, if he wishes to postpone the Project in lieu of so terminating the Consulting Engineer's appointment, require the Consulting Engineer in writing to suspend the carrying out of his services under this Agreement for the time being. Upon such a termination the Client shall pay to the Consulting Engineer a sum calculated in accordance with the provisions of clause 16.1.'

2.	Clause 3.1 shall be deleted, and the following substituted:

	'The copyright in all drawings, reports, specifications, bills of quantities, calculations and other documents provided by the Consulting Engineer in connection with the Project shall remain vested in the Consulting Engineer. The Consulting Engineer shall promptly supply the Client with conveniently reproducible copies of all such drawings and other documents, as and when they are intended to be issued or used. The Client shall have a licence to copy and use such drawings and other documents for any purpose related to the execution, maintenance, repair, renewal, letting, promotion, advertisement and/or extension of the Project. Save as aforesaid, the Client shall not make copies of such drawings or other documents nor shall he use the same in connection with any other works without the prior written approval of the Consulting Engineer, such approval not to be unreasonably withheld or delayed.'

3.	In clause 5.1, after 'Consulting Engineer' where it first appears, there shall be added 'has exercised and.'

4.	After clause 5.2, there shall be added:

	'5.3	Without prejudice to the generality of clause 5.1, the Consulting Engineer warrants:

	5.3.1	that he has not specified and will not specify for use;

	5.3.2	that he has exercised and will continue to exercise reasonable skill, care and diligence to see that there are not used;

	5.3.3	that he is not aware and has no reason to suspect or believe that there have been or will be used; and

5.3.4   that he will promptly notify the Client in writing if he becomes aware or has reason to suspect or believe that there have been or will be used;

in or in connection with the Project, any of the materials or substances identified in clause 5.4.

5.4   The said materials or substances are:

5.4.1   high alumina cement in structural elements;

5.4.2   wood wool slabs in permanent formwork to concrete;

5.4.3   calcium chloride in admixtures for use in reinforced concrete;

5.4.4   asbestos products;

5.4.5   naturally occurring aggregates for use in reinforced concrete which do not comply with British Standard 882: 1983 and/or naturally occurring aggregates for use in concrete which do not comply with British Standard 8110: 1985.'

5.   Clause 13.1 (g) shall be deleted.

6.   Clause 14.2 shall be deleted.

7.   Clauses 16.1 (c), 16.2 (c) and 16.3 (b) shall be deleted.

8.   In clause 18.1 (c), delete 'as if purchased new.'

9.   The following clauses shall be added:

'20.   **Professional Indemnity Insurance**∗

20.1   The Consulting Engineer shall maintain professional indemnity insurance covering (inter alia) all liability hereunder upon customary and usual terms and conditions prevailing for the time being in the insurance market, and with reputable insurers lawfully carrying on such insurance business in the United Kingdom, in an amount of not less than   pounds (£   ) for any one occurrence or series of occurrences arising out of any one event for a period beginning now and ending 15 (fifteen) years after the date of practical completion of the Project for the purposes of the building contract, provided always that such insurance is available at commercially reasonable rates. The said terms and conditions shall not include any term or condition to the effect that the Consulting Engineer must discharge any liability before being entitled to recover from the insurers, or any other term or condition which might adversely affect the rights of any person to recover from the insurers pursuant to the Third Parties (Rights Against Insurers) Act 1930, or any amendment or re-enactment thereof. The Consulting Engineer shall not, without the prior approval in writing of the Client, settle or compromise with the insurers any claim which the Consulting Engineer may have against the insurers and which relates to a claim by the Client against the Consulting Engineer, or by any act or omission lose or prejudice the Consulting Engineer's right to make or proceed with such a claim against the insurers.

20.2   Any increased or additional premium required by insurers by reason of the Consulting Engineer's own claims record or other acts, omissions, matters or things particular to the Consulting Engineer shall be deemed to be within commercially reasonable rates.

20.3   The Consulting Engineer shall immediately inform the Client if such insurance ceases to be available at commercially reasonable rates in order that the Consulting Engineer and the Client can discuss means of best protecting the respective positions of the Client and the Consulting Engineer in respect of the Project in the absence of such insurance.

20.4   The Consulting Engineer shall fully co-operate with any measures reasonably required by the Client, including (without limitation) completing any proposals for insurance and associated documents, maintaining such insurance at rates above commercially reasonable rates if the Client undertakes in writing to reimburse the Consulting Engineer in respect of the net cost of such insurance to the Consulting Engineer above commercially reasonable rates or, if the Client effects such insurance at rates at or above commercially reasonable rates, reimbursing the Client in respect of what the net cost of such insurance to the Client would have been at commercially reasonable rates.

20.5   As and when reasonably requested to do so by the Client, the Consulting Engineer shall produce for inspection documentary evidence (including, if required by the Client, the original of the relevant insurance documents) that his professional indemnity insurance is being maintained.

20.6   The above obligations in respect of professional indemnity insurance shall continue notwithstanding termination of this Agreement for any reason whatsoever, including (without limitation) breach by the Client.

21.   Collateral Warranties✱

21.1   The Consulting Engineer shall within 7 (seven) working days of the Client's request so to do:

21.1.1   execute, in favour of any persons who have entered or shall enter into an agreement for the provision of finance in connection with the Project, a Deed in the form annexed as Annexe 'A', or a similar form reasonably required by the Client, and deliver the same to the Client, together with a guarantee in the form annexed as Annexe 'C', or a similar form reasonably required by the Client, from [   ] and [   ] **OR** [the ultimate parent company of the Consulting Engineer, namely [   ] [Limited] **OR** [PLC]], in respect of the Consulting Engineer's obligations pursuant to such Deed; and

21.1.2   execute, in favour of any persons who have acquired or shall acquire any interest in or over the Project or any part thereof, a Deed in the form annexed as Annexe 'B', or a similar form reasonably required by the Client, and deliver the same to the Client, together with a guarantee in the form annexed as Annexe 'C', or a similar form reasonably required by the Client, from [   ] and [   ] **OR** [the ultimate parent company of the Consulting Engineer, namely [   ] [Limited] **OR** [PLC]], in respect of the Consulting Engineer's obligations pursuant to such Deed.

21.2   The above obligation for the provision of Deeds and guarantees in favour of third parties shall continue notwithstanding termination of this Agreement for any reason whatsoever, including (without limitation) breach by the Client. However, any such Deed given after such termination shall be amended by the Client so as to refer to the fact and date of such termination, to omit any obligation to continue to exercise skill, care and diligence after such termination, and to omit any provision enabling a third party to assume the position of the Client.

22.   **[Personal Guarantees] OR [Parent Company Guarantee]**✱

Upon execution of this Deed, the Consulting Engineer shall deliver to the Client a guarantee from [   ] and [   ] **OR** parent company guarantee from its ultimate parent company, namely [   ][Limited] **OR** [PLC], for its performance of this Deed in the form annexed as Annexe 'D'.'

COMMENTARY

Memorandum of Agreement
Articles 3, 4 and 5

The relevant Articles set out elaborate provisions concerning fees, etc., which will have to be negotiated and completed in each case.

Clause 20: Professional Indemnity Insurance

See commentary upon clause 9 of Form 6.

Clause 21: Collateral Warranties

For Annexes 'A', 'B' and 'C', see Forms 6, 7, 8 and 9.

Clause 22: Personal Guarantees or Parent Company Guarantee

For Annexe 'D', see Forms 10 and 11.

FORM 4: BUILDING SERVICES ENGINEER'S APPOINTMENT, INCORPORATING ACE AGREEMENT 4A

DATED _____ **199**

[   ]

('the Client')

-and-

[   ]

('the Consulting Engineer')

BUILDING SERVICES ENGINEER'S APPOINTMENT

for professional services
relating to the construction of
[   ]

✳ An asterisk indicates that there is a relevant note in the commentary upon this Form.

THIS AGREEMENT made the   day of   One thousand nine hundred and ninety-

BETWEEN:

(1)

[of] **OR** [whose registered office is at]

('the Client'); and

(2)

[of] **OR** [whose registered office is at]

('the Consulting Engineer');

INCORPORATES the recitals and Articles 1, 2, 3, 4 and 5 of the Memorandum of Agreement and the Conditions of Engagement included in The Association of Consulting Engineers' 'Conditions of Engagement – Agreement 4A – For Engineering Services in Relation to Sub-Contract Works' (Full Duties), as printed in March 1990, and as amended in August 1990 and by the Schedule.

IN WITNESS whereof the parties have executed this Deed in duplicate on the date first stated above.

SCHEDULE

Recitals to the Memorandum of Agreement

The name and address of the Architect are

The Project is the construction of

Memorandum of Agreement✻

Articles 3, 4 and 5 of the Memorandum of Agreement shall be completed as follows:

Amendments to Conditions of Engagement

1.   Clause 2.3 shall be deleted, and the following substituted:

'If at any time the Client so decides, he may by notice in writing to the Consulting Engineer terminate the Consulting Engineer's appointment under this Agreement, provided that the Client may, if he wishes to postpone the Project in lieu of so terminating the Consulting Engineer's appointment, require the Consulting Engineer in writing to suspend the carrying out of his services under this Agreement for the time being. Upon such a termination the Client shall pay to the Consulting Engineer a sum calculated in accordance with the provisions of clause 21.1.'

2.   Clause 3.1 shall be deleted, and the following substituted:

'The copyright in all drawings, reports, specifications, bills of quantities, calculations and other documents provided by the Consulting Engineer in connection with the Project shall remain vested in the Consulting Engineer. The Consulting Engineer shall promptly supply the Client with conveniently reproducible copies of all such drawings and other documents, as and when they are intended to be issued or used. The Client shall have a licence to copy and use such drawings and other documents for any purpose related to the execution, maintenance, repair, renewal, letting, promotion, advertisement and/or extension of the Project. Save as aforesaid, the Client shall not make copies of such drawings or other documents nor shall he use the same in connection with any other works without the prior written approval of the Consulting Engineer, such approval not to be unreasonably withheld or delayed.'

3.   In clause 5.1, after 'Consulting Engineer' where it first appears, there shall be added 'has exercised and.'

4.   After clause 5.3, there shall be added:

'5.4   Without prejudice to the generality of clause 5.1, the Consulting Engineer warrants:

5.4.1   that he has not specified and will not specify for use;

5.4.2   that he has exercised and will continue to exercise reasonable skill, care and diligence to see that there are not used;

5.4.3   that he is not aware and has no reason to suspect or believe that there have been or will be used; and

5.4.4   that he will promptly notify the Client in writing if he becomes aware or has reason to suspect or believe that there have been or will be used;

in or in connection with the Project, any of the materials or substances identified in clause 5.5.

5.5   The said materials or substances are:

5.5.1   high alumina cement in structural elements;

5.5.2   wood wool slabs in permanent formwork to concrete;

5.5.3   calcium chloride in admixtures for use in reinforced concrete;

5.5.4   asbestos products;

5.5.5   naturally occurring aggregates for use in reinforced concrete which do not comply with British Standard 882: 1983 and/or naturally occurring aggregates for use in concrete which do not comply with British Standard 8110: 1985.'

5.   Clause 18.1 (g) shall be deleted.

6.   Clause 19.2 shall be deleted.

7.   Clauses 21.1 (c), 21.2 (b) and 21.3 (b) shall be deleted.

8.   In clause 23.1 (f), delete 'as if purchased new.'

9.   The following clauses shall be added:

'25.   **Professional Indemnity Insurance✼**

25.1   The Consulting Engineer shall maintain professional indemnity insurance covering (inter alia) all liability hereunder upon customary and usual terms and conditions prevailing for the time being in the insurance market, and with reputable insurers lawfully carrying on such insurance business in the United Kingdom, in an amount of not less than   pounds (£   ) for any one occurrence or series of occurrences arising out of any one event for a period beginning now and ending 15 (fifteen) years after the date of practical completion of the Project for the purposes of the building contract, provided always that such insurance is available at commercially reasonable rates. The said terms and conditions shall not include any term or condition to the effect that the Consulting Engineer must discharge any liability before being entitled to recover from the insurers, or any other term or condition which might adversely affect the rights of any person to recover from the insurers pursuant to the Third Parties (Rights Against Insurers) Act 1930, or any amendment or re-enactment thereof. The Consulting Engineer shall not, without the prior approval in writing of the Client, settle or compromise with the insurers any claim which the Consulting Engineer may have against the insurers and which relates to a claim by the Client against the Consulting Engineer, or by any act or omission lose or prejudice the Consulting Engineer's right to make or proceed with such a claim against the insurers.

25.2   Any increased or additional premium required by insurers by reason of the Consulting Engineer's own claims record or other acts, omissions, matters or things particular to the Consulting Engineer shall be deemed to be within commercially reasonable rates.

25.3   The Consulting Engineer shall immediately inform the Client if such insurance ceases to be available at commercially reasonable rates in order that the Consulting Engineer and the Client can discuss means of best protecting the respective positions of the Client and the Consulting Engineer in respect of the Project in the absence of such insurance.

25.4   The Consulting Engineer shall fully co-operate with any measures reasonably required by the Client, including (without limitation) completing any proposals for insurance and associated documents, maintaining such insurance at rates above commercially reasonable rates if the Client undertakes in writing to reimburse the Consulting Engineer in respect of the net cost of such insurance to the Consulting Engineer above commercially reasonable rates or, if the Client effects such insurance at rates at or above commercially reasonable rates, reimbursing the Client in respect of what the net cost of such insurance to the Client would have been at commercially reasonable rates.

25.5   As and when reasonably requested to do so by the Client, the Consulting Engineer shall produce for inspection documentary evidence (including, if required by the Client, the original of the relevant insurance documents) that his professional indemnity insurance is being maintained.

25.6   The above obligations in respect of professional indemnity insurance shall continue notwithstanding termination of this Agreement for any reason whatsoever, including (without limitation) breach by the Client.

26.   Collateral Warranties✻

26.1   The Consulting Engineer shall within 7 (seven) working days of the Client's request so to do:

26.1.1 execute, in favour of any persons who have entered or shall enter into an agreement for the provision of finance in connection with the Project, a Deed in the form annexed as Annexe 'A', or a similar form reasonably required by the Client, and deliver the same to the Client, together with a guarantee in the form annexed as Annexe 'C', or a similar form reasonably required by the Client, from [   ] and [   ] **OR** [the ultimate parent company of the Consulting Engineer, namely [   ] [Limited] **OR** [PLC]], in respect of the Consulting Engineer's obligations pursuant to such Deed; and

26.1.2 execute, in favour of any persons who have acquired or shall acquire any interest in or over the Project or any part thereof, a Deed in the form annexed as Annexe 'B', or a similar form reasonably required by the Client, and deliver the same to the Client, together with a guarantee in the form annexed as Annexe 'C', or a similar form reasonably required by the Client, from [   ] and [   ] **OR** [the ultimate parent company of the Consulting Engineer, namely [   ] [Limited] **OR** [PLC]], in respect of the Consulting Engineer's obligations pursuant to such Deed.

26.2 The above obligation for the provision of Deeds and guarantees in favour of third parties shall continue notwithstanding termination of this Agreement for any reason whatsoever, including (without limitation) breach by the Client. However, any such Deed given after such termination shall be amended by the Client so as to refer to the fact and date of such termination, to omit any obligation to continue to exercise skill, care and diligence after such termination, and to omit any provision enabling a third party to assume the position of the Client.

27. **[Personal Guarantees] OR [Parent Company Guarantee]***

Upon execution of this Deed, the Consulting Engineer shall deliver to the Client a guarantee from [   ] and [   ] **OR** parent company guarantee from its parent company, namely [   ] [Limited] **OR** [PLC], for its performance of this Deed in the form annexed as Annexe 'D'.'

COMMENTARY

Memorandum of Agreement
Articles 3, 4 and 5

The relevant Articles set out elaborate provisions concerning fees, etc., which will have to be negotiated and completed in each case.

Clause 25: Professional Indemnity Insurance

See commentary upon clause 9 of Form 6.

Clause 26: Collateral Warranties

For Annexes 'A', 'B' and 'C', see Forms 6, 7, 8 and 9.

Clause 27: Personal Guarantees or Parent Company Guarantees

For Annexe 'D', see Forms 10 and 11.

FORM 5: PROJECT MANAGER'S APPOINTMENT, INCORPORATING THE RICS PROJECT MANAGER'S APPOINTMENT

DATED _____ **199**

[   ]

('the Client')

-and-

[   ]

('the Project Manager')

PROJECT MANAGER'S APPOINTMENT

for professional services
relating to the construction of
[   ]

✱ An asterisk indicates that there is a relevant note in the commentary upon this Form.

THIS AGREEMENT made the   day of   One thousand nine hundred and ninety-

BETWEEN:

(1)

    [of] **OR** [whose registered office is at]

    ('the Client'); and

(2)

    [of] **OR** [whose registered office is at]

    ('the Project Manager');

INCORPORATES the Project Management Agreement and Conditions of Engagement issued by the Royal Institution of Chartered Surveyors and the Project Management Association of the RICS✱ in May 1992, as amended by the Schedule.

IN WITNESS whereof the parties have executed this Deed in duplicate on the date first stated above.

SCHEDULE

Memorandum of Agreement
Recital A

The Project is the construction of

Clause 5

Add to clause 5(a):

    'or approved or acquiesced in by him without reasonable skill, care and diligence.'

Add to clauses 5(b) and (c):

'unless approved or acquiesced in by the Project Manager without reasonable skill, care and diligence.'

Clause 6

Clause 6 shall be deleted, and the following substituted:

'6.1   The Project Manager has performed and will perform the Services with reasonable skill, care and diligence.

6.2   Without prejudice to the generality of clause 6.1, the Project Manager warrants:

6.2.1   that he has not specified and will not specify for use;

6.2.2   that he has exercised and will continue to exercise reasonable skill, care and diligence to see that there are not used;

6.2.3   that he is not aware and has no reason to suspect or believe that there have been or will be used; and

6.2.4   that he will promptly notify the Client in writing if he becomes aware or has reason to suspect or believe that there have been or will be used;

in or in connection with the Project, any of the materials or substances identified in clause 6.3.

6.3   The said materials or substances are:

6.3.1   high alumina cement in structural elements;

6.3.2   wood wool slabs in permanent formwork to concrete;

6.3.3   calcium chloride in admixtures for use in reinforced concrete;

6.3.4   asbestos products;

6.3.5   naturally occurring aggregates for use in reinforced concrete which do not comply with British Standard 882: 1983 and/or naturally occurring aggregates for use in concrete which do not comply with British Standard 8110: 1985.'

Clause 8

The date to be inserted is   199   .

Clause 9

The period of delay to be inserted is [12 (twelve) months].

Clause 10(a)

The fee to be inserted is   pounds (£   ).

Clause 10(c)

The date to be inserted is   199   .

Clause 10(d)

The rate of interest to be inserted is   % (per cent).

Clause 11

The name to be inserted is

Clause 12✱

Clause 12 shall be deleted.

Clause 13✱

Clause 13 shall be deleted.

Conditions of Engagement

Clause 4: Collateral Warranty✱

Clause 4 shall be deleted, and the following substituted:

'4.1   The Project Manager shall within 7 (seven) working days of the Client's request so to do:

4.1.1   execute, in favour of any persons who have entered or shall enter into an agreement for the provision of finance in connection with the Project, a Deed in the form annexed as Annexe 'A', or a similar form reasonably required by the Client, and deliver the same to the Client, together with a guarantee in the form annexed as Annexe 'C', or a similar form reasonably required by the Client, from [   ] and [   ] **OR** [the ultimate parent company of the Project Manager, namely [   ] [Limited] **OR** [PLC]], in respect of the Project Manager's obligations pursuant to such Deed; and

4.1.2   execute, in favour of any persons who have acquired or shall acquire any interest in or over the Project or any part thereof, a Deed in the form annexed as Annexe 'B', or a similar form reasonably required by the Client, and deliver the same to the Client, together with a guarantee in the form annexed as Annexe 'C', or a similar form reasonably required by the Client, from [   ] and [   ] **OR** [the ultimate parent company of the Project Manager, namely [   ] [Limited] **OR** [PLC]], in respect of the Project Manager's obligations pursuant to such Deed.

4.2   The above obligation for the provision of Deeds and guarantees in favour of third parties shall continue notwithstanding termination of this Agreement for any reason whatsoever, including (without limitation) breach by the Client. However, any such Deed given after such termination shall be amended by the Client so as to refer to the fact and date of such termination, to omit any obligation to continue to exercise skill, care and diligence after such termination, and to omit any provision enabling a third party to assume the position of the Client.'

Clause 5: Duration of Engagement

Add to clause 5.2:

'such consent not to be unreasonably withheld or delayed.'

Clause 6: The Obligations of the Client

The last sentence of clause 6.3 shall be deleted, and the following substituted:

> 'The Client shall procure that such Consultants shall provide information reasonably required by the Project Manager in the performance of his duties when reasonably required, without charge to the Project Manager.'

Clause 15: Professional Indemnity Insurance*

Clause 15 shall be deleted, and the following substituted:

'15.1   The Project Manager shall maintain professional indemnity insurance covering (inter alia) all liability hereunder upon customary and usual terms and conditions prevailing for the time being in the insurance market, and with reputable insurers lawfully carrying on such insurance business in the United Kingdom, in an amount of not less than   pounds (£   ) for any one occurrence or series of occurrences arising out of any one event for a period beginning now and ending 15 (fifteen) years after the date of practical completion of the Project for the purposes of the building contract, provided always that such insurance is available at commercially reasonable rates. The said terms and conditions shall not include any term or condition to the effect that the Project Manager must discharge any liability before being entitled to recover from the insurers, or any other term or condition which might adversely affect the rights of any person to recover from the insurers pursuant to the Third Parties (Rights Against Insurers) Act 1930, or any amendment or re-enactment thereof. The Project Manager shall not, without the prior approval in writing of the Client, settle or compromise with the insurers any claim which the Project Manager may have against the insurers and which relates to a claim by the Client against the Project Manager, or by any act or omission lose or prejudice the Project Manager's right to make or proceed with such a claim against the insurers.

15.2   Any increased or additional premium required by insurers by reason of the Project Manager's own claims record or other acts, omissions, matters or things particular to the Project Manager shall be deemed to be within commercially reasonable rates.

15.3   The Project Manager shall immediately inform the Client if such insurance ceases to be available at commercially reasonable rates in order that the Project Manager and the Client can discuss means of best protecting the respective positions of the Client and the Project Manager in respect of the Project in the absence of such insurance.

15.4   The Project Manager shall fully co-operate with any measures reasonably required by the Client, including (without limitation) completing any proposals for insurance and associated documents, maintaining such insurance at rates above commercially reasonable rates if the Client undertakes in writing to reimburse the Project Manager in respect of the net cost of such insurance to the Project Manager above commercially reasonable rates or, if the Client effects such insurance at rates at or above commercially reasonable rates, reimbursing the Client in respect of what the net cost of such insurance to the Client would have been at commercially reasonable rates.

15.5   As and when reasonably requested to do so by the Client, the Project Manager shall produce for inspection documentary evidence (including, if required by the Client, the original of the relevant insurance documents) that his professional indemnity insurance is being maintained.'

15.6   The above obligations in respect of professional indemnity insurance shall continue notwithstanding termination of this Agreement for any reason whatsoever, including (without limitation) breach by the Client.'

Clause 16: Copyright

Clause 16 shall be deleted, and the following substituted:

'The copyright in all documents, calculations and data in whatever form provided by the Project Manager in connection with the Project shall remain vested in the Project Manager. The Project Manager shall promptly supply to the Client conveniently reproducible copies of all such documents, calculations and data, as and when they are intended to be issued or used. The Client shall have a licence to copy and use such drawings and other documents for any purpose related to the execution, maintenance, repair, renewal, letting, promotion, advertisement and/or extension of the Project.'

The following clause shall be added:

'Clause 18: [Personal Guarantees] OR [Parent Company Guarantee]*

Upon execution of this Deed, the Project Manager shall deliver to the Client a guarantee from [   ] and [   ] OR parent company guarantee from its ultimate parent company, namely [   ] [Limited] OR [PLC], for its performance of this Deed in the form annexed as Annexe 'D'.

APPENDIX [A] [B] [C] [D] [E] [F]✱

COMMENTARY

RICS Project Management Agreement and Conditions of Engagement

A very useful RICS Guidance Note is also published, particularly in relation to the completion of Appendices A-F.

Memorandum of Agreement

Clause 12: Professional Indemnity Insurance

See clause 15 of the Conditions of Engagement, as substituted.

Clause 13

This clause has been deleted because it will almost invariably operate solely in favour of one party, the Project Manager. It is unlikely that the Client will be deemed to have assumed implied obligations under this document.

Conditions of Engagement

Clause 4: Collateral Warranty

For Annexes 'A', 'B' and 'C', see Forms 6, 7, 8 and 9.

Clause 15: Professional Indemnity Insurance

See commentary upon clause 9 of Form 6.

Clause 18: Personal Guarantees or Parent Company Guarantee

For Annexe 'D', see Forms 10 and 11.

Appendices A-F

The published Appendices A-F should be completed and attached. If the Project Manager is to be a Construction Manager in a construction management project, that duty could easily be added to Appendix A.

FORM 6: COLLATERAL WARRANTY BY CONSULTANT TO FUNDING INSTITUTION, FOR USE AS AN ANNEXE TO FORMS 1–5

ANNEXE 'A'

DATED _____ **199**

[   ]

('the Firm')

-and-

[   ]

('the Client')

-and-

[   ]

('the Company')

COLLATERAL WARRANTY BY CONSULTANT
TO FUNDING INSTITUTION

in respect of professional services
relating to the construction of

[   ]

✶ An asterisk indicates that there is a relevant note in the commentary upon this Form.

THIS AGREEMENT✻ is made the   day of   One thousand nine hundred and ninety-

BETWEEN:

(1)

[of] **OR** [whose registered office is at]

('the Firm');

(2)

[of] **OR** [whose registered office is at]

('the Client'); and

(3)

[whose registered office is at]

('the Company', which term shall include its successors and assigns).✻

WHEREAS:

(A)   The Company has entered into an agreement ('the Finance Agreement') with the Client for the provision of certain finance in connection with the carrying out of a project of development briefly described as
at

('the Development').

[The Company entered into the Finance Agreement, and enters into this Agreement, on its own behalf and as agent for a syndicate of banks. Each of the banks which are members of the syndicate from time to time, including banks joining the syndicate after the date of this Agreement, shall be entitled to the benefit of this Agreement in addition to the Company.✻]

(B)✻   By a contract ('the Appointment') dated   199   (a copy of which is annexed and signed for identification purposes by the parties) the Client has appointed the Firm as [architects] **OR** [consulting structural engineers] **OR** [consulting building services engineers] **OR** [surveyors] **OR** [project managers] in connection with the Development.

(C)   The Client has entered or may enter into a building contract ('the Building Contract') with [   ] [Limited] **OR** [PLC] **OR** [a building contractor to be selected by the Client] for the construction of the Development.

NOW in consideration of £1 (one pound) paid by the Company to the Firm (receipt of which the Firm hereby acknowledges) **THIS DEED WITNESSETH** as follows:

1.✱   The Firm warrants that it has exercised and will continue to exercise reasonable skill, care and diligence in the performance of its duties to the Client under the Appointment.

2.1✱   Without prejudice to the generality of clause 1, the Firm further warrants:

      2.1.1   that it has not specified and will not specify for use;

      2.1.2   that it has exercised and will continue to exercise reasonable skill, care and diligence to see that there are not used;

      2.1.3   that it is not aware and has no reason to suspect or believe that there have been or will be used; and

      2.1.4   that it will promptly notify the Company in writing if it becomes aware or has reason to suspect or believe that there have been or will be used;

in or in connection with the Development, any of the materials or substances identified in clause 2.2.

2.2   The said materials or substances are:

      2.2.1   high alumina cement in structural elements;

      2.2.2   wood wool slabs in permanent formwork to concrete;

      2.2.3   calcium chloride in admixtures for use in reinforced concrete;

      2.2.4   asbestos products;

      2.2.5   naturally occurring aggregates for use in reinforced concrete which do not comply with British Standard 882: 1983 and/or naturally occurring aggregates for use in concrete which do not comply with British Standard 8110: 1985.

3.   The Company has no authority to issue any direction or instruction to the Firm in relation to performance of the Firm's duties under the Appointment unless and until

the Company has given notice under clauses 5 or 6.

4.   The Firm acknowledges that the Client has paid all fees and expenses due and owing to the Firm under the Appointment up to the date of this Agreement. The Company has no liability to the Firm in respect of fees and expenses under the Appointment unless and until the Company has given notice under clauses 5 or 6.

5.✳   The Firm agrees that, in the event of the termination of the Finance Agreement by the Company, or the occurrence of an event of default under the Finance Agreement, the Firm will, if so required by notice in writing given by the Company and subject to clause 7, accept the instructions of the Company or its appointee to the exclusion of the Client in respect of the Development upon the terms and conditions of the Appointment. The Client acknowledges that the Firm shall be entitled to rely on a notice given to the Firm by the Company under this clause 5 as conclusive evidence for the purposes of this Agreement of the termination of the Finance Agreement by the Company, or the occurrence of any such event of default.

6.✳   The Firm further agrees that it will not, without first giving the Company not less than 21 (twenty-one) days' notice in writing, exercise any right it may have to terminate the Appointment, or to treat the same as having been repudiated by the Client, or discontinue the performance of any duties to be performed by the Firm pursuant thereto for any reason whatsoever (including, without limitation, any act or omission by or on behalf of the Client). The Firm's right to terminate the Appointment with the Client, or treat the same as having been repudiated, or discontinue performance shall cease if, within such period of notice and subject to clause 7, the Company shall give notice in writing to the Firm requiring the Firm to accept the instructions of the Company or its appointee to the exclusion of the Client in respect of the Development upon the terms and conditions of the Appointment.

7.✳   It shall be a condition of any notice given by the Company under clauses 5 or 6 that the Company or its appointee accepts liability for payment of the fees payable to the Firm under the Appointment and for performance of the Client's obligations under the Appointment, including payment of any fees outstanding at the date of such notice, but excluding any such fees in respect of which funds were advanced to the Client pursuant to the Finance Agreement before the date of such notice. Upon the issue of any notice by the Company under clauses 5 or 6, the Appointment shall continue in full force and effect as if no right of termination on the part of the Firm had arisen and the Firm shall be liable to the Company or its appointee under the Appointment in lieu of its liability to the Client. If any notice given by the Company under clauses 5 or 6 requires the Firm to accept the instructions of the Company's appointee, the Company shall be liable to the Firm as guarantor for the payment of all sums from time to time due to the Firm from the Company's appointee.

8.✱   The copyright in all drawings, reports, specifications, bills of quantities, calculations and other similar documents provided by the Firm in connection with the Development shall remain vested in the Firm, but the Company and its appointee shall have a licence to copy and use such drawings and other documents, and to reproduce the designs contained in them, for any purpose related to the Development including, but without limitation, the construction, completion, maintenance, letting, promotion, advertisement, reinstatement, repair and/or extension of the Development. The Firm shall, if the Company so requests and undertakes in writing to pay the Firm's reasonable copying charges, promptly supply the Company with conveniently reproducible copies of all such drawings and other documents.

9.1✱   The Firm shall maintain professional indemnity insurance covering (inter alia) all liability hereunder upon customary and usual terms and conditions prevailing for the time being in the insurance market, and with reputable insurers lawfully carrying on such insurance business in the United Kingdom, in an amount of not less than   pounds (£   ) for any one occurrence or series of occurrences arising out of any one event for a period beginning now and ending 15 (fifteen) years after the date of practical completion of the Development for the purposes of the Building Contract, provided always that such insurance is available at commercially reasonable rates. The said terms and conditions shall not include any term or condition to the effect that the Firm must discharge any liability before being entitled to recover from the insurers, or any other term or condition which might adversely affect the rights of any person to recover from the insurers pursuant to the Third Parties (Rights Against Insurers) Act 1930, or any amendment or re-enactment thereof. The Firm shall not, without the prior approval in writing of the Company, settle or compromise with the insurers any claim which the Firm may have against the insurers and which relates to a claim by the Company against the Firm or by any act or omission lose or prejudice the Firm's right to make or proceed with such a claim against the insurers.

9.2   Any increased or additional premium required by insurers by reason of the Firm's own claims record or other acts, omissions, matters or things particular to the Firm shall be deemed to be within commercially reasonable rates.

9.3   The Firm shall immediately inform the Company if such insurance ceases to be available at commercially reasonable rates in order that the Firm and the Company can discuss means of best protecting the respective positions of the Company and the Firm in respect of the Development in the absence of such insurance.

9.4   The Firm shall fully co-operate with any measures reasonably required by the Company, including (without limitation) completing any proposals for insurance and associated documents, maintaining such insurance at rates above commercially reasonable rates if the Company undertakes in writing to reimburse the Firm in respect of the net cost of such insurance to the Firm above commercially reasonable rates or, if the Company effects such insurance at rates at or above commercially reasonable rates, reimbursing the Company in respect of what the net cost of such insurance to the Company would have been at commercially reasonable rates.

9.5   As and when it is reasonably requested to do so by the Company or its appointee under clauses 5 or 6, the Firm shall produce for inspection documentary evidence (including, if required by the Company or such appointee, the original of the relevant insurance documents) that its professional indemnity insurance is being maintained.

10.✱   The Client has agreed to be a party to this Agreement for the purpose of clause 12 and for acknowledging that the Firm shall not be in breach of the Appointment by complying with the obligations imposed on it by this Agreement.

11.✱   This Agreement may be assigned by the Company and its successors and assigns without the consent of the Client or the Firm being required.

12.✱   The Client and the Firm undertake with the Company not to vary, or depart from, the terms and conditions of the Appointment without the prior written consent of the Company, and agree that no such variation or departure made without such consent shall be binding on the Company, or affect or prejudice the Company's rights hereunder, or under the Appointment, or in any other way.

13.   Any notice to be given by the Firm hereunder shall be deemed to be duly given if it is delivered by hand at or sent by registered post or recorded delivery to the Company at its registered office, and any notice to be given by the Company hereunder shall be deemed to be duly given if it is addressed to ['the Senior Partner'] **OR** ['the Managing Director'] and delivered by hand at or sent by registered post or recorded delivery to the above-mentioned address of the Firm or to the principal business address of the Firm for the time being and, in the case of any such notices, the same shall if sent by registered post or recorded delivery be deemed to have been received forty-eight hours after being posted.

IN WITNESS whereof the Client and the Firm have executed this Deed on the date first stated above.

COMMENTARY

In order to facilitate acceptance by consultants and their professional indemnity insurers, this Form is closely based upon BPF Form CoWa/F, Second Edition 1990 (Appendix B). See Chapter 3. Material changes from CoWa/F are indicated in the commentary. Neither CoWa/F nor this Form contain provisions, such as warranties of fitness for purpose and compliance with contractual performance specifications and requirements, which would go substantially beyond the normal professional obligations and duties of 'the Firm'. 'The Company' should be advised that the basic obligation of 'the Firm' is to exercise due care, not to warrant a result. If 'the Company' wishes to impose such warranties, strong resistance may be expected from 'the Firm' and its professional indemnity insurers.

'The Company'

'Successors and assigns' has been substituted for the CoWa/F expression 'all permitted assignees under this Agreement'.

Recital (A)

The additional recital in square brackets has been added to CoWa/F. The banking lawyer concerned should ensure that this recital is accurate and effective, and conforms with the finance documents.

Recital (B)

No material changes from CoWa/F, except that it is provided that a copy of the Appointment shall be annexed. This is in order to record exactly what it is, and to enable 'the Company' to consider its terms. See also clause 12.

Operative provisions

Clause 1: Duty of Care

Clause 1 of CoWa/F gives options between 'skill and care' and 'skill, care and diligence', which the form states should reflect the Appointment. However, regardless of the terms of the Appointment, it is submitted that 'the Firm' should undertake 'diligence' to 'the Company'. The proviso to clause 1 of CoWa/F 'provided that the Firm shall have no greater liability to the Company by virtue of this Agreement than it would have had if the Company had been named as a joint client under the Appointment' has been deleted. It is submitted that the rights of 'the Company' hereunder should stand alone, and not be affected by what its rights might theoretically have been if it had been a 'joint client under the Appointment'.

Clause 2: Proscribed Materials

This clause is quite different from clause 2 of CoWa/F, under which 'the Firm' warrants 'that it has exercised and will continue to exercise reasonable skill and care to see that, unless authorized by the Client in writing or, where such authorization is given orally, confirmed by the Firm to the Client in writing, none of the following has been or will be specified by the Firm for use in the construction of those parts of the Development to which the Appointment relates...'. Further proscribed materials may, of course, be added, if so advised by an expert on building materials. The list included here is the same as that in CoWa/F.

Clause 5: 'Step-in' Rights

The banking lawyer concerned should ensure that this clause conforms with the finance documents. In relation to the Finance Agreement, events of default are referred to, as well as termination.

Clause 6: 'Step-in' Rights

The first sentence of clause 6 of CoWa/F has been amended in order to make it clear that 'the Firm' is not to discontinue its performance for any reason whatsoever, without giving notice to 'the Company'. Some standard professional terms of engagement, e.g. those of the RIBA, enable the consultant to resign from the engagement on reasonable notice for any reason. Therefore, the right to take over such an engagement may be of little useful effect. The forms of consultants' appointments given in this book do not permit the consultants to resign at will.

Clause 7: Guarantee to Consultant

The concluding works of the first sentence have been added to clause 7 of CoWa/F. In the circumstances contemplated, 'the Client' may have drawn down funds for the payment of fees, but become insolvent before paying them to 'the Firm'. In the absence of the added words, 'the Company' might then have to pay the same sum again.

Clause 8: Copyright

The licence contained in the first sentence of clause 8 of CoWa/F has been enlarged to cover the reproduction of designs contained in the relevant documents for the extension of the Development. Clause 7 of the Scottish equivalent of CoWa/F, the Royal Incorporation of Architects in Scotland Duty of Care Agreement 1988, gives 'the Company' such a right, unlike CoWa/F. As a result of the amended licence, the second and third sentences of clause 8 of CoWa/F have been deleted. The third sentence would otherwise exclude the Firm's liability for use of the relevant documents for extensions. The last sentence has been added to clause 8 of CoWa/F. In the circumstances contemplated, e.g. the insolvency of 'the Client', 'the Company' may find it difficult to obtain copies except from 'the Firm'.

Clause 9: Professional Indemnity Insurance

Several amendments have been made to clause 9 of CoWa/F. The effect of the amendments is as follows:

- the insurance cannot be on unusual terms, e.g. with wide exclusions or very large excesses;
- it must be with insurers carrying on business in the UK, and therefore complying with UK insurance legislation;
- a period from the date of the Deed to 15 years after practical completion is given (see Limitation Act 1980, section 14B, inserted by Latent Damage Act 1986, section 1) but a period of 12 or 6 years is more common in practice;
- the penultimate sentence of clause 9.1 is intended to avoid the effect of *Firma C-Trade SA v Newcastle Protection and Indemnity Insurance Association (The Fanti)* and *Socony Mobil Oil Co. Inc. v West of England Ship Owners Mutual Insurance Association (London) Limited (The Padre Island)* [1990] 2 All ER 705, HL, in which it was held that a clause such as that specified could, in the event of the insured's insolvency, defeat third party rights under the Third Parties (Rights Against Insurers) Act 1930;
- the last sentence of clause 9.1 is intended to prevent 'the Firm' settling claims on disadvantageous terms, which it might otherwise be free to do (see *Normid Housing Association Limited v Ralphs* (1988) 43 BLR 19);
- any loading of premiums particular to 'the Firm', as opposed to general increases in premium levels, must be borne by 'the Firm';
- if premiums do escalate unduly, 'the Company' may require any reasonable measures to be taken, and 'the Firm' must co-operate and continue to bear the cost of commercially reasonable premiums; and

– it is made clear that actual policy documents must be produced if required, so enabling 'the Company' to check that the insurance is upon customary and usual terms, and complies with this clause.

Clause 10: Client as a Party

The concluding words ('this Agreement') have been substituted for 'Clauses 5 and 6' in clause 10 of CoWa/F. 'The Firm' now also undertakes relevant obligations under clause 2.

Clause 11: Assignment

This clause is quite different from the restrictive clause 11 of CoWa/F, which reads:

'This Agreement may be assigned by the Company by way of absolute legal assignment to another company providing finance or re-finance in connection with the Development without the consent of the Client or the Firm being required and such assignment shall be effective upon written notice thereof being given to the Client and to the Firm.'

The clause as re-drafted only provides for assignment in entirety. Purchasers and tenants of part of the Development would normally expect to receive collateral warranties directly from 'the Firm'.

Clause 12: Variation of the Appointment

This clause is quite different, and has a different (and self-explanatory) purpose, from clause 12 of CoWa/F, which reads:

'The Client undertakes to the Firm that warranty agreements in the Model Form CoWa/F published by the British Property Federation or in substantially similar form have been or will be entered into between [   ] on the one hand and the Company on the other hand.'

Clause 12 of CoWa/F is no doubt designed to enable 'the Firm' and its professional indemnity insurers to benefit from the Civil Liability (Contribution) Act 1978. However, the CoWa/F clause would be difficult to operate in practice, and it is submitted that the arrangements agreed with other consultants are not the legitimate concern of 'the Firm'. Clause 12 is, of course, in the interests of 'the Company', but not of 'the Client' or 'the Firm'.

FORM 7: COLLATERAL WARRANTY BY CONSULTANT TO ACQUIRER, FOR USE AS AN ANNEXE TO FORMS 1–5

ANNEXE 'B'

DATED _____ **199**

[   ]

('the Firm')

-and-

[   ]

('the Client')

-and-

[   ]

('the Acquirer')

COLLATERAL WARRANTY BY CONSULTANT TO ACQUIRER

in respect of professional services
relating to the construction of
[   ]

✱ **An asterisk indicates that there is a relevant note in the commentary upon this Form.**

THIS AGREEMENT* is made the   day of   One thousand nine hundred and ninety-

BETWEEN:

(1)

      [of] **OR** [whose registered office is at]

      ('the Firm'); and

(2)

      [of] **OR** [whose registered office is at]

      ('the Client'); and

(3)

      [of] **OR** [whose registered office is at]

      ('the Acquirer', which term shall include its successors and assigns).

WHEREAS:

(A)   The Acquirer intends to acquire, or has acquired, an interest in a project of development briefly described as
at

                                              ('the Development').

(B)   By a contract ('the Appointment') dated   199   (a copy of which is annexed and signed for identification purposes by the parties) the Client has appointed the Firm as [architects] **OR** [consulting structural engineers] **OR** [consulting building services engineers] **OR** [surveyors] **OR** [project managers] in connection with the Development.

(C)   The Client has entered or may enter into a building contract ('the Building Contract') with [   ] [Limited] **OR** [PLC] **OR** [a building contractor to be selected by the Client] for the construction of the Development.

NOW in consideration of £1 (one pound) paid by the Acquirer to the Firm (receipt of which the Firm hereby acknowledges) **THIS DEED WITNESSETH** as follows:

1. The Firm warrants that it has exercised and will continue to exercise reasonable skill, care and diligence in the performance of its duties to the Client under the Appointment.

2.1 Without prejudice to the generality of clause 1, the Firm further warrants:

    2.1.1 that it has not specified and will not specify for use;

    2.1.2 that it has exercised and will continue to exercise reasonable skill, care and diligence to see that there are not used;

    2.1.3 that it is not aware and has no reason to suspect or believe that there have been or will be used; and

    2.1.4 that it will promptly notify the Acquirer in writing if it becomes aware or has reason to suspect or believe that there have been or will be used;

in or in connection with the Development, any of the materials or substances identified in clause 2.2.

2.2 The said materials or substances are:

    2.2.1 high alumina cement in structural elements;

    2.2.2 wood wool slabs in permanent formwork to concrete;

    2.2.3 calcium chloride in admixtures for use in reinforced concrete;

    2.2.4 asbestos products;

    2.2.5 naturally occurring aggregates for use in reinforced concrete which do not comply with British Standard 882: 1983 and/or naturally occurring aggregates for use in concrete which do not comply with British Standard 8110: 1985.

3. The Acquirer has no authority to issue any direction or instruction to the Firm in relation to performance of the Firm's duties under the Appointment.

4. The Firm acknowledges that the Client has paid all fees and expenses due and owing to the Firm under the Appointment up to the date of this Agreement. The Acquirer has no liability to the Firm in respect of fees and expenses under the Appointment.

5. The copyright in all drawings, reports, specifications, bills of quantities, calculations and other similar documents provided by the Firm in connection with the Development

shall remain vested in the Firm, but the Acquirer and its appointee shall have a licence to copy and use such drawings and other documents, and to reproduce the designs contained in them, for any purpose related to the Development including, but without limitation, the construction, completion, maintenance, letting, promotion, advertisement, reinstatement, repair and/or extension of the Development. The Firm shall, if the Acquirer so requests and undertakes in writing to pay the Firm's reasonable copying charges, promptly supply the Acquirer with conveniently reproducible copies of all such drawings and other documents.

6.1   The Firm shall maintain professional indemnity insurance covering (inter alia) all liability hereunder upon customary and usual terms and conditions prevailing for the time being in the insurance market, and with reputable insurers lawfully carrying on such insurance business in the United Kingdom, in an amount of not less than   pounds (£   ) for any one occurrence or series of occurrences arising out of any one event for a period beginning now and ending 15 (fifteen) years after the date of practical completion of the Development for the purposes of the Building Contract, provided always that such insurance is available at commercially reasonable rates. The said terms and conditions shall not include any term or condition to the effect that the Firm must discharge any liability before being entitled to recover from the insurers, or any other term or condition which might adversely affect the rights of any person to recover from the insurers pursuant to the Third Parties (Rights Against Insurers) Act 1930, or any amendment or re-enactment thereof. The Firm shall not, without the prior approval in writing of the Acquirer, settle or compromise with the insurers any claim which the Firm may have against the insurers and which relates to a claim by the Acquirer against the Firm, or by any act or omission lose or prejudice the Firm's right to make or proceed with such claim against the insurers.

6.2   Any increased or additional premium required by insurers by reason of the Firm's own claims record or other acts, omissions, matters or things particular to the Firm shall be deemed to be within commercially reasonable rates.

6.3   The Firm shall immediately inform the Acquirer if such insurance ceases to be available at commercially reasonable rates in order that the Firm and the Acquirer can discuss means of best protecting the respective positions of the Acquirer and the Firm in respect of the Development in the absence of such insurance.

6.4   The Firm shall fully co-operate with any measures reasonably required by the Acquirer, including (without limitation) completing any proposals for insurance and associated documents, maintaining such insurance at rates above commercially reasonable rates if the Acquirer undertakes in writing to reimburse the Firm in respect of the net cost of such insurance to the Firm above commercially reasonable rates or, if the Acquirer effects such insurance at rates at or above commercially reasonable rates, reimbursing the Acquirer in respect of what the net cost of such insurance to the Acquirer would have been at commercially reasonable rates.

6.5   As and when it is reasonably requested to do so by the Acquirer the Firm shall produce for inspection documentary evidence (including, if required by the Acquirer, the original of the relevant insurance documents) that its professional indemnity insurance is being maintained.

7.   The Client has agreed to be a party to this Agreement for the purpose of clause 9 and for acknowledging that the Firm shall not be in breach of the Appointment by complying with the obligations imposed on it by this Agreement.

8.   This Agreement may be assigned by the Acquirer and its successors and assigns without the consent of the Client or the Firm being required.

9.   The Client and the Firm undertake with the Acquirer not to vary, or depart from, the terms and conditions of the Appointment without the prior written consent of the Acquirer, and agree that no such variation or departure made without such consent shall be binding on the Acquirer, or affect or prejudice the Acquirer's rights hereunder, or in any other way.

10.   Any notice to be given by the Firm hereunder shall be deemed to be duly given if it is delivered by hand at or sent by registered post or recorded delivery to the above-mentioned address of the Acquirer or to the principal business address of the Acquirer for the time being, and any notice to be given by the Acquirer hereunder shall be deemed to be duly given if it is addressed to ['the Senior Partner'] **OR** ['the Managing Director'] and delivered by hand at or sent by registered post or recorded delivery to the above-mentioned address of the Firm or to the principal business address of the Firm for the time being and, in the case of any such notices, the same shall if sent by registered post or recorded delivery be deemed to have been received forty-eight hours after being posted.

IN WITNESS whereof the Client and the Firm have executed this Deed on the date first stated above.

COMMENTARY

This Form closely resembles Form 6, except that clauses 5, 6 and 7 of that Form, which enable the funding institution to take over the relevant consultant's appointment as substitute client, are omitted. Reference should be made to the commentary upon that Form.

FORM 8: PERSONAL GUARANTEE FOR THE PERFORMANCE OF CONSULTANT'S COLLATERAL WARRANTY, FOR USE AS AN ANNEXE TO FORMS 1–5

ANNEXE 'C'

DATED _____ **199**

[   ] and [   ]

('the Guarantors')

-and-

[   ]

('the Fund')

OR

[   ]

('the Acquirer')

GUARANTEE

in respect of a Consultant's Collateral Warranty
relating to the construction of
[   ]

✳ An asterisk indicates that there is a relevant note in the commentary upon this Form.

THIS AGREEMENT is made the   day of   One thousand nine hundred and ninety-

BETWEEN:

(1)

[of]

('the Guarantors'); and

(2)

[of] **OR** [whose registered office is at]

(['the Fund'] **OR** ['the Acquirer'], which term shall include its successors and assigns).

WHEREAS by an Agreement ('the Collateral Warranty') dated   199 and made between [   ] ('the Firm') [   ] (as Client) and the [Fund] **OR** [Acquirer], the Firm assumed certain obligations towards the [Fund] **OR** [Acquirer].

NOW THIS DEED WITNESSETH that if the Firm defaults in the discharge of any of the Firm's obligations under or pursuant to the Collateral Warranty, the Guarantors will [jointly and severally]✳ indemnify the [Fund] **OR** [Acquirer] against all loss and damage thereby caused to the [Fund] **OR** [Acquirer], and no alterations in the Collateral Warranty, and no extension of time, forbearance or forgiveness, nor any act, matter or thing whatsoever except an express release by Deed by the [Fund] **OR** [Acquirer], shall in any way release the Guarantors from any liability hereunder.

IN WITNESS whereof the Guarantors have executed this Deed on the date first stated above.

COMMENTARY

See paragraphs 23 and 24, Chapter 1.

FORM 9: PARENT COMPANY GUARANTEE FOR THE PERFORMANCE OF CONSULTANT'S COLLATERAL WARRANTY, FOR USE AS AN ANNEXE TO FORMS 1–5

ANNEXE 'C'

DATED _____ **199**

[   ]

('the Guarantor')

-and-

[   ]

('the Fund')

OR

[   ]

('the Acquirer')

GUARANTEE

in respect of a Consultant's Collateral Warranty
relating to the construction of
[   ]

✻ An asterisk indicates that there is a relevant note in the commentary upon this Form.

THIS AGREEMENT✻ is made the   day of   One thousand nine hundred and ninety-

BETWEEN:

(1)

[whose registered office is at]

('the Guarantor'); and

(2)

[of] **OR** [whose registered office is at]

(['the Fund'] **OR** ['the Acquirer'], which term shall include its successors and assigns).

WHEREAS by an Agreement ('the Collateral Warranty') dated   199 and made between [   ] ('the Firm') [   ] (as Client) and the [Fund] **OR** [Acquirer], the Firm assumed certain obligations towards the [Fund] **OR** [Acquirer].

NOW THIS DEED WITNESSETH that if the Firm defaults in the discharge of any of the Firm's obligations under or pursuant to the Collateral Warranty, the Guarantor will indemnify the [Fund] **OR** [Acquirer] against all loss and damage thereby caused to the [Fund] **OR** [Acquirer], and no alterations in the Collateral Warranty, and no extension of time, forbearance or forgiveness, nor any act, matter or thing whatsoever except an express release by Deed by the [Fund] **OR** [Acquirer], shall in any way release the Guarantor from any liability hereunder.

IN WITNESS whereof the Guarantor has executed this Deed on the date first stated above.

COMMENTARY

See paragraph 23, Chapter 1.

FORM 10: PERSONAL GUARANTEE FOR THE PERFORMANCE OF CONSULTANT'S APPOINTMENT, FOR USE AS AN ANNEXE TO FORMS 1–5

ANNEXE 'D'

DATED _____ **199**

[   ]

('the Guarantors')

-and-

[   ]

('the Client')

GUARANTEE

in respect of professional services
relating to the construction of

[   ]

✱ An asterisk indicates that there is a relevant note in the commentary upon this Form.

THIS AGREEMENT is made the   day of   One thousand nine hundred and ninety-

BETWEEN:

(1)

[of]

('the Guarantors'); and

(2)

[of] **OR** [whose registered office is at]

('the Client', which term shall include its successors and assigns).

WHEREAS by a contract ('the Appointment') dated   199   the Client has appointed [   ] [Limited] **OR** [PLC] ('the Firm') as [architects] **OR** [consulting structural engineers] **OR** [consulting building services engineers] **OR** [surveyors] **OR** [project managers] in connection with the carrying out of a project of development briefly described as [   ].

NOW THIS DEED WITNESSETH that if the Firm defaults in the discharge of any of the Firm's obligations under or pursuant to the Appointment, the Guarantors will [jointly and severally]✱ indemnify the Client against all loss and damage thereby caused to the Client, and no alterations in the Appointment, and no extension of time, forbearance or forgiveness, nor any act, matter or thing whatsoever except an express release by Deed by the Client, shall in any way release the Guarantors from any liability hereunder.

IN WITNESS whereof the Guarantors have executed this Deed on the date first stated above.

COMMENTARY

See paragraph 24, Chapter 1.

FORM 11: PARENT COMPANY GUARANTEE FOR THE PERFORMANCE OF CONSULTANT'S APPOINTMENT, FOR USE AS AN ANNEXE TO FORMS 1–5

ANNEXE 'D'

DATED _____ **199**

[   ]

('the Guarantor')

-and-

[   ]

('the Client')

GUARANTEE

in respect of professional services
relating to the construction of
[   ]

✻ An asterisk indicates that there is a relevant note in the commentary upon this Form.

THIS AGREEMENT✳ is made the   day of   One thousand nine hundred and ninety-

BETWEEN:

(1)

[whose registered office is at]

('the Guarantor'); and

(2)

[of] **OR** [whose registered office is at]

('the Client', which term shall include its successors and assigns).

WHEREAS by a contract ('the Appointment') dated   199   the Client has appointed [   ] [Limited] **OR** [PLC] ('the Firm') as [architects] **OR** [consulting structural engineers] **OR** [consulting building services engineers] **OR** [surveyors] **OR** [project managers] in connection with the carrying out of a project of development briefly described as [   ].

NOW THIS DEED WITNESSETH that if the Firm defaults in the discharge of any of the Firm's obligations under or pursuant to the Appointment, the Guarantor will indemnify the Client against all loss and damage thereby caused to the Client, and no alterations in the Appointment, and no extension of time, forbearance or forgiveness, nor any act, matter or thing whatsoever except an express release by Deed by the Client, shall in any way release the Guarantor from any liability hereunder.

IN WITNESS whereof the Guarantor has executed this Deed on the date first stated above.

COMMENTARY

See paragraph 25, Chapter 1.

FORM 12: MAIN CONTRACT INCORPORATING JCT STANDARD FORM OF BUILDING CONTRACT, 1980 EDITION, PRIVATE WITH QUANTITIES, WITH OPTIONS FOR SECTIONAL COMPLETION

DATED _____ **199**

[   ]

('the Employer')

-and-

[   ]

('the Contractor')

MAIN CONTRACT

for the construction of

[   ]

✱ An asterisk indicates that there is a relevant note in the commentary upon this Form.

THIS AGREEMENT made the   day of   One thousand nine hundred and ninety-

BETWEEN:

(1)

[of] **OR** [whose registered office is at]

('the Employer'); and

(2)

[of] **OR** [whose registered office is at]

('the Contractor');

INCORPORATES the Joint Contracts Tribunal Standard Form of Building Contract, 1980 Edition, Private With Quantities, as printed in   199   , incorporating Amendments 1 (January 1984), 2 (November 1986), 4 (July 1987), 5 (January 1988), 6 (July 1988), 7 (July 1988), 8 (April 1989), 9 (July 1990), 10 (March 1991), and 11 (July 1992, as corrected in September 1992), [as modified by the JCT Sectional Completion Supplement (1981 Edition, revised   199   )], and as amended by the Schedule.

IN WITNESS whereof the parties have executed this Deed in duplicate on the date first stated above.

SCHEDULE

Recitals

First recital

The Works are the construction of

at

The Drawings and Bills of Quantities were prepared by or under the direction of [the Architect].

Third recital

The Contract Drawings are numbered:

Fourth recital

Delete 'the Finance (No. 2) Act 1975' and substitute 'Part XIII (Miscellaneous Special Provisions) Chapter IV (Sub-Contractors in the Construction Industry) of the Income and Corporation Taxes Act 1988.'

Articles of Agreement

Article 2: Contract Sum

The sum to be inserted is   pounds (£   ).

Article 3: Architect✱

The name and address of the Architect are

[of] **OR** [whose registered office is at]

Delete 'not being a person to whom the Contractor no later than 7 days after such nomination shall object for reasons considered to be sufficient by an Arbitrator appointed in accordance with article 5'.

Article 4: Quantity Surveyor✱

The name and address of the Quantity Surveyor are

[of] **OR** [whose registered office is at]

Delete 'not being a person to whom the Contractor no later than 7 days after such nomination shall object for reasons considered to be sufficient by an Arbitrator appointed in accordance with article 5.'

Amendments to Conditions

Clause 1.3: Definitions

The definition of 'Conditions' shall include any additional clauses or provisions hereby added, and the Conditions as hereby amended and added to.

Add 'his successors and assigns' to the definition of 'Employer'.✳

The definition of 'Joint Names Policy' shall be deleted, and the following substituted:

'Joint Names Policy: a policy of insurance which includes the Contractor and the Employer and such other persons as the Employer may reasonably require (including, without limitation, any persons who have entered or shall enter into an agreement for the provision of finance in connection with the Works, and any persons who have acquired or shall acquire any interest in or over the Works or any part thereof) as the insured.'

Clause 2.2.1: Contract Bills – relation to Articles, Conditions and Appendix✳

Delete.

Clauses 19.1.1 and 19.1.2: Assignment✳

Delete clause 19.1.1 and the first sentence of clause 19.1.2, and substitute:

'19.1.1 The Contractor shall not, without the written consent of the Employer, assign this Contract.

19.1.2 Where clause 19.1.2 is stated in the Appendix to apply, the Employer may assign this Contract.'

Clause 24: Damages for non-completion✳

In clause 24.2.1, delete 'as the Employer may require in writing not later than the date of the Final Certificate'.

Clause 25.4: Relevant Events✳

[Clause 25.4.7 shall be deleted.]

Clause 25.4.10 shall be deleted.

Add to clause 25.4 as an additional sub-clause:

'any breach of this Contract by the Employer or any act or omission on the part of the Employer, the Architect or the Quantity Surveyor'.

Clause 27: Determination by Employer✳

Clause 27.2.4 shall be deleted.

The last sentence of clause 27.5.3 shall be deleted.

The proviso to clause 27.6.4.1 shall be deleted.

[Clause 27.7 shall be deleted.]

OR

[In clause 27.7.1, delete '6 months', and substitute '12 (twelve) months'.

In clause 27.7.2, delete '6 month', and substitute '12 (twelve) month'.]

Clause 28A: Determination by Employer or Contractor✳

Clause 28A.5.5 shall be deleted.

Clause 30: Certificates and Payments✳

In clause 30.1.1.2, delete:

'Notwithstanding the fiduciary interest of the Employer in the Retention as stated in clause 30.5.1...'.

Clause 30.5.1 shall be deleted, and the following substituted:

'30.5.1 the Employer's interest in the Retention shall not be fiduciary, either as trustee for the Contractor or any Nominated Sub-Contractor or any other person, or in any other capacity; the relationship of the Employer and the Contractor with regard to the Retention shall be solely that of debtor and unsecured creditor, subject to the terms hereof; and the Employer shall have no obligation to invest the Retention or any part thereof;'

Clause 30.5.3 shall be deleted, and the following substituted:

'30.5.3 the Employer shall have no obligation to segregate the Retention or any part thereof in a separate banking account, or in any other manner whatsoever; and shall be entitled to the full beneficial interest in the Retention and every part thereof (and, without limitation, interest thereon and income arising therefrom) unless and until the Retention is paid to the Contractor pursuant to this Contract.'

Clause 30.3 shall be deleted.

In clause 30.9.2, the last word ('either') of the first paragraph shall be deleted.

In clause 30.9.2.1, the last word ('or') shall be deleted.

Clause 30.9.2.2 shall be deleted.

The concluding words of clause 30.9.2 ('whichever shall be the earlier') shall be deleted.

Clause 31: Finance (No.2) Act 1975 – statutory tax deduction scheme

References to 'the Act', 'the Regulations' or any legislation shall be interpreted to include references to any amendment or re-enactment thereof.

[Clause 35: Nominated Sub-Contractors✳

Clause 35.18.2 shall be deleted.

In clause 35.21, the word 'not' shall be deleted where it occurs for the second time.

Clause 35.24.9 shall be deleted, and the following substituted:

'35.24.9 The amount properly payable to the Nominated Sub-Contractor under the Sub-Contract resulting from such further nomination shall be included in the amount stated as due in Interim Certificates and added to the Contract Sum. Provided that any extra amount, payable by the Employer in respect of the Sub-Contractor nominated under the further nomination over the price of the Nominated Sub-Contractor who first entered into a sub-contract in respect of the relevant sub-contract works, resulting from such further nomination, may at the time or at any time after such amount is certified in respect of the Sub-Contractor nominated under the further nomination, be deducted by the Employer from monies due or to become due to the Contractor under this Contract, or may be recoverable from the Contractor by the Employer as a debt.'

Clause 38.4: Provisions relating to clause 38✻

Clause 38.4.8.1 shall be deleted.

Clause 39.5: Provisions relating to clause 39✻

Clause 39.5.8.1 shall be deleted.

Clause 40: Use of price adjustment formulae✻

Clause 40.7.2.1 shall be deleted.

Additional Conditions

The following clauses shall be added:

'42:   Collateral Warranties✻

42.1   The Contractor shall within 7 (seven) working days of the Employer's request so to do:

42.1.1   execute, in favour of any persons who have entered or shall enter into an agreement for the provision of finance in connection with the Works, a Deed in the form annexed as Annexe 'A', or a similar form reasonably required by the Employer, and deliver the same to the Employer, together with a guarantee in the form annexed as Annexe 'C', or a similar form reasonably required by the Employer, from the ultimate parent company of the Contractor, namely [   ] [Limited] **OR** [PLC], in respect of the Contractor's obligations pursuant to such Deed; and

42.1.2   execute, in favour of any persons who have acquired or shall acquire any interest in or over the Works or any part thereof, a Deed in the form annexed as Annexe 'B', or a similar form reasonably required by the Employer, and deliver the same to the Employer, together with a guarantee in the form annexed as Annexe 'C', or a similar form reasonably required by the Employer, from the ultimate parent company of the Contractor, namely [   ] [Limited] **OR** [PLC], in respect of the Contractor's obligations pursuant to such Deed.

42.2   The Contractor shall (without request) use reasonable endeavours to procure that each sub-contract or supply contract with each Nominated Sub-Contractor or Supplier, and any other sub-contractor or supplier (if requested by the Employer), shall contain obligations on the relevant sub-contractor or supplier to execute, in favour of the Employer and/or (within 7 (seven) working days of the Employer's request so to do) any persons who have entered or shall enter into an agreement for the provision of finance in connection with the Works and/or in favour of any persons who have acquired or shall acquire any interest in or over the Works or any part thereof, a Deed in the form annexed as Annexe 'D', or a similar form reasonably required by the Employer, and deliver the same to the Employer; together in each case (if requested by the Employer) with a guarantee in the form annexed as Annexe 'E', or a similar form reasonably required by the Employer, from the ultimate parent company of the relevant sub-contractor or supplier in respect of the sub-contractor's or supplier's obligations pursuant to such Deed. The Contractor shall enforce such obligations, or such modified obligations as are referred to below.

42.3   If, despite the Contractor having used such reasonable endeavours, the sub-contractor or supplier will not accept such obligations, or will only accept them in a modified form, the Contractor shall notify the Employer, who may agree in writing that the relevant sub-contract or supply contract need not contain such obligations, or that the relevant obligations may be in a modified form agreeable to the sub-contractor or supplier.

42.4   Failing such agreement by the Employer in the case of a proposed Nominated Sub-Contractor or Supplier, the Contractor shall not be obliged to enter into a sub-contract or supply contract with that Nominated Sub-Contractor or Supplier.

42.5   Failing such agreement by the Employer in the case of any other sub-contractor or supplier, the Contractor shall not enter into a relevant sub-contract or supply contract with that sub-contractor or supplier.

42.6 The above obligations for the provision of Deeds and guarantees in favour of third parties shall continue notwithstanding termination of this Contract, or determination of the Contractor's employment hereunder, in either case for any reason whatsoever, including (without limitation) breach by the Employer. However, any such Deed given after such termination or determination shall be amended by the Employer so as to refer to the fact and date of such termination or determination, to omit any obligation to continue to exercise skill, care or diligence or fulfil obligations pursuant to this Contract after such termination or determination, and to omit any provision enabling a third party to assume the position of employer of the Contractor.

43: **[Performance Bond] OR [Parent Company Guarantee]✻**

Upon execution of this Contract, the Contractor shall deliver to the Employer a performance bond from [   ] [Limited] **OR** [PLC] in an amount of 10% (ten per cent) of the Contract Sum in the form annexed as Annexe 'F' **OR** parent company guarantee from its ultimate parent company, namely [   ] [Limited] **OR** [PLC], for its performance of this Contract in the form annexed as Annexe 'F'.

44: **Proscribed Materials**

44.1 The Contractor warrants:

44.1.1 that he has not used or specified and will not use or specify for use;

44.1.2 that he has exercised and will continue to exercise reasonable skill, care and diligence to see that there are not used;

44.1.3 that he is not aware and has no reason to suspect or believe that there have been or will be used; and

44.1.4 that he will promptly notify the Employer in writing if he becomes aware or has reason to suspect or believe that there have been or will be used;

in or in connection with the Works, any of the materials or substances identified in clause 44.2.

44.2 The said materials or substances are:

44.2.1   high alumina cement in structural elements;

44.2.2   wood wool slabs in permanent formwork to concrete;

44.2.3   calcium chloride in admixtures for use in reinforced concrete;

44.2.4   asbestos products;

44.2.5   naturally occurring aggregates for use in reinforced concrete which do not comply with British Standard 882: 1983 and/or naturally occurring aggregates for use in concrete which do not comply with British Standard 8110: 1985.'

APPENDIX*

COMMENTARY

Article 3: Architect
Article 4: Quantity Surveyor

See paragraph 5, Chapter 2.

Clause 1.3: Definitions

See paragraph 6, Chapter 2.

Clause 2.2.1: Contract Bills – relation to Articles, Conditions and Appendix

See paragraph 7, Chapter 2.

Clauses 19.1.1 and 19.1.2: Assignment

See paragraphs 8 and 9, Chapter 2.

Clause 24: Damages for non-completion

See paragraph 10, Chapter 2.

Clause 25.4: Relevant Events

See paragraphs 11, 12 and 13, Chapter 2. With regard to the deletion of clause 25.4.7, see the commentary upon clause 35 below.

Clause 27: Determination by Employer

See paragraphs 15-20, Chapter 2.

Clause 28A: Determination by Employer or Contractor

See paragraph 21, Chapter 2.

Clause 30: Certificates and Payments

See paragraphs 22, 23 and 24, Chapter 2.

Clause 35: Nominated Sub-Contractors

The Contractor's responsibilities to the Employer for Nominated Sub-Contractors are strictly limited:

– he is not responsible for their delay (clause 25.4.7);
– he is not responsible for their performance of the matters to which clause 2.1 of Agreement NSC/W relates (which are the design of the Sub-Contract Works by the Nominated Sub-Contractor, the **selection** of materials and goods for the Sub-Contract Works by the Nominated Sub-Contractor, and the satisfaction of any relevant performance specification or requirement) but he remains responsible for the **supply** of workmanship, materials and goods by the Nominated Sub-Contractors (clause 35.21); and
– in the event of the default or insolvency of a Nominated Sub-Contractor, the adverse financial consequences effectively fall on the Employer, not the Contractor (clause 35.24.9).

The Employer may, of course, avoid these consequences by refraining from nominating sub-contractors, if possible.

The Form contains optional amendments in order to negative the disadvantages to the Employer of nomination, but strong resistance from Contractors may be expected.

See also paragraph 26, Chapter 1.

Clause 38.4: Provisions relating to clause 38
Clause 39.5: Provisions relating to clause 39
Clause 40: Use of price adjustment formulae

See paragraph 25, Chapter 2.

Clause 42: Collateral Warranties

For Annexes 'A', 'B', 'C', 'D' and 'E', see Forms 19, 20, 28, 21 and 29.

Clause 43: Performance Bond or Parent Company Guarantee

For Annexes 'F', see Forms 30 and 31.

Appendix

The published Appendix should be completed and attached. With regard to the entry re: clause 19.1.2, delete 'of benefits after Practical Completion' or, if the JCT Sectional Completion Supplement is being used, the equivalent words in the Appendix (Sectional Completion Supplement). Periods of delay have to be entered opposite the Appendix items relating to clauses 28 and 28A. Clause 28.2.2 enables the Contractor to determine his employment if the Works are suspended for the relevant period by reason of certain factors which can be laid at the door of the Employer, such as Architect's instructions for variations of the Works, delay in the issue of necessary instructions, and delay in the execution of work by the Employer or other contractors. Under clause 28A, either the Employer or the Contractor may determine the employment of the Contractor, if the Works are suspended for the relevant period by reason of a number of 'neutral' events. These provisions are reasonable enough in principle, but the JCT suggests, at p.13 of JCT Amendment 11, that the relevant periods of delay should be only one or three months. It is suggested that these periods are too short, from the Employer's point of view, and should be, say, six months. Where the relevant delay is the Employer's responsibility, the Contractor will normally be compensated by a loss and expense claim under clause 26. It is suggested that the appointor of arbitrators should be the President or a Vice-President of the Chartered Institute of Arbitrators, in view of that Institute's training of its arbitrators.

FORM 13: MAIN CONTRACT INCORPORATING JCT STANDARD FORM OF BUILDING CONTRACT, 1980 EDITION, PRIVATE WITH APPROXIMATE QUANTITITES, WITH OPTIONS FOR SECTIONAL COMPLETION

DATED _____ **199**

[   ]

('the Employer')

-and-

[   ]

('the Contractor')

MAIN CONTRACT

for the construction of

[   ]

✱ An asterisk indicates that there is a relevant note in the commentary upon this Form.

THIS AGREEMENT made the   day of   One thousand nine hundred and ninety-

BETWEEN:

(1)

    [of] **OR** [whose registered office is at]

    ('the Employer'); and

(2)

    [of] **OR** [whose registered office is at]

    ('the Contractor');

INCORPORATES the Joint Contracts Tribunal Standard Form of Building Contract, 1980 Edition, Private With Approximate Quantities, as printed in   199 , incorporating Amendments 1 (January 1984), 2 (November 1986), 4 (July 1987), 5 (January 1988), 6 (July 1988), 7 (July 1988), 8 (April 1989), 9 (July 1990), 10 (March 1991) and 11 (July 1992, as corrected in September 1992), [as modified by the JCT Sectional Completion Supplement (1981 Edition, revised   199)], and as amended by the Schedule.

IN WITNESS whereof the parties have executed this Deed in duplicate on the date first stated above.

SCHEDULE

Recitals

First recital

The Works are the construction of

at

The Drawings and Bills of Approximate Quantities were prepared by or under the direction of [the Architect].

Second recital

The Tender Price is   pounds (£   ).

Third recital

The Contract Drawings are numbered:

Fourth recital

Delete 'the Finance (No. 2) Act 1975' and substitute 'Part XIII (Miscellaneous Special Provisions) Chapter IV (Sub-Contractors in the Construction Industry) of the Income and Corporation Taxes Act 1988.'

Articles of Agreement

Article 3: Architect

The name and address of the Architect are

[of] **OR** [whose registered office is at]

Delete 'not being a person to whom the Contractor no later than 7 days after such nomination shall object for reasons considered to be sufficient by an Arbitrator appointed in accordance with article 5'.

Article 4: Quantity Surveyor

The name and address of the Quantity Surveyor are

[of] **OR** [whose registered office is at]

Delete 'not being a person to whom the Contractor no later than 7 days after such nomination shall object for reasons considered to be sufficient by an Arbitrator appointed in accordance with article 5.'

Amendments to Conditions✱

COMMENTARY

Continue as in Form 12, and see the commentary upon that Form.

However, clause 43 should refer to 'Tender Price,' rather than 'Contract Sum'.

FORM 14: MAIN CONTRACT INCORPORATING JCT STANDARD FORM OF BUILDING CONTRACT, 1980 EDITION, PRIVATE WITHOUT QUANTITIES

DATED   **199**

[   ]

('the Employer')

-and-

[   ]

('the Contractor')

MAIN CONTRACT

for the construction of

[   ]

✱ An asterisk indicates that there is a relevant note in the commentary upon this Form.

THIS AGREEMENT made the   day of   One thousand nine hundred and ninety-

BETWEEN:

(1)

[of] **OR** [whose registered office is at]

('the Employer'); and

(2)

[of] **OR** [whose registered office is at]

('the Contractor');

INCORPORATES the Joint Contracts Tribunal Standard Form of Building Contract, 1980 Edition, Private Without Quantities, as printed in   199   , incorporating Amendments 1 (January 1984), 2 (November 1986), 3 (March 1987), 4 (July 1987), 5 (January 1988), 6 (July 1988), 8 (April 1989), 9 (July 1990), 10 (March 1991) and 11 (July 1992, as corrected in September 1992), and as amended by the Schedule.

IN WITNESS whereof the parties have executed this Deed in duplicate on the date first stated above.

SCHEDULE

Recitals

First recital

The Works are the construction of

at

The Contract Drawings are numbered:

The Contract Drawings and the [Specification] **OR** [Schedules of Work] were prepared by or under the direction of [the Architect].

Second recital

[Alternative A shall apply.]

OR

[Alternative B shall apply. The Contractor has supplied the Employer with a [Contract Sum Analysis as defined in Clause 1.3] **OR** [Schedule of Rates on which the Contract Sum is based].]

Third recital

Delete 'the Finance (No. 2) Act 1975' and substitute 'Part XIII (Miscellaneous Special Provisions) Chapter IV (Sub-Contractors in the Construction Industry) of the Income and Corporation Taxes Act 1988.'

Articles of Agreement

Article 2: Contract Sum

The sum to be inserted is   pounds (£   ).

Article 3: Architect

The name and address of the Architect are

[of] **OR** [whose registered office is at]

Delete 'not being a person to whom the Contractor no later than 7 days after such nomination shall object for reasons considered to be sufficient by an Arbitrator appointed in accordance with article 5'.

[Article 4A: Quantity Surveyor

The name and address of the Quantity Surveyor are

[of] **OR** [whose registered office is at]

Delete 'not being a person to whom the Contractor no later than 7 days after such nomination shall object for reasons considered to be sufficient by an Arbitrator appointed in accordance with article 5.']

OR

[Article 4B: Exercise of functions of Quantity Surveyor

The functions ascribed by the conditions to the 'Quantity Surveyor' shall be exercised by

[of] **OR** [whose registered office is at]

Delete 'not being a person to whom the Contractor no later than 7 days after such nomination shall object for reasons considered to be sufficient by an Arbitrator appointed in accordance with article 5.']

Amendments to Conditions✱

COMMENTARY

Continue as in Form 12, and see the commentary upon that Form, but the title of clause 2.2.1 in JCT 80 Without Quantities is:

'Specification/Schedules of Work – relation to Articles, Conditions and Appendix'

FORM 15: MAIN CONTRACT INCORPORATING JCT STANDARD FORM OF BUILDING CONTRACT WITH CONTRACTOR'S DESIGN, 1981 EDITION, INCLUDING COLLATERAL WARRANTY BY CONSULTANT EMPLOYED BY 'DESIGN AND BUILD' CONTRACTOR, PERFORMANCE BOND AND PARENT COMPANY GUARANTEE

DATED   **199**

[   ]

('the Employer')

-and-

[   ]

('the Contractor')

MAIN CONTRACT

for the design and construction of

[   ]

✱ **An asterisk indicates that there is a relevant note in the commentary upon this Form.**

THIS AGREEMENT made the   day of   One thousand nine hundred and ninety-

BETWEEN:

(1)

[of] **OR** [whose registered office is at]

('the Employer'); and

(2)

[of] **OR** [whose registered office is at]

('the Contractor');

INCORPORATES the Joint Contracts Tribunal Standard Form of Building Contract With Contractor's Design, 1981 Edition, as printed in   199   , incorporating Amendments 1 (November 1986), 2 (July 1987), 3 (February 1988), 4 (July 1988), 5 (April 1989) and 6 (November 1990) and, if stated in Appendix 1, the Supplementary Provisions issued in February 1988 which modify the Conditions, and as amended by the Schedule.

IN WITNESS whereof the parties have executed this Deed in duplicate on the date first stated above.

SCHEDULE

Recitals

First recital

The Works are the design and construction of

at

Fourth recital

Delete 'the Finance (No. 2) Act 1975' and substitute 'Part XIII (Miscellaneous Special Provisions) Chapter IV (Sub-Contractors in the Construction Industry) of the Income and Corporation Taxes Act 1988.'

Articles of Agreement

Article 2: Contract Sum

The sum to be inserted is   pounds (£   ).

Article 3: Employer's Agent

The name and address of the Employer's Agent are

[of] **OR** [whose registered office is at]

Amendments to Conditions

Clause 1.3: Definitions

The definition of 'Conditions' shall include any additional clauses or provisions hereby added, and the Conditions as hereby amended and added to.

Add 'his successors and assigns' to the definition of 'Employer'.✱

The definition of 'Joint Names Policy' shall be deleted, and the following substituted:

'Joint Names Policy: a policy of insurance which includes the Contractor and the Employer and such other persons as the Employer may reasonably require (including, without limitation, any persons who have entered or shall enter into an agreement for the provision of finance in connection with the Works, and any persons who have acquired or shall acquire any interest in or over the Works or any part thereof) as the insured.'

**Clause 2.2: Employer's Requirements etc. –
relation to Articles, Conditions and Appendices✳**

Delete.

[Clause 2.5.1: Contractor's design warranty✳]

Clause 5.6.2: Limits to use of documents

Add 'construction, completion, extension' after 'advertisement'.

Clause 18.1.1 and 18.1.2: Assignment✳

Delete clause 18.1.1 and the first sentence of clause 18.1.2, and substitute:

'18.1.1   The Contractor shall not, without the written consent of the Employer, assign this Contract.

18.1.2   Where clause 18.1.2 is stated in Appendix 1 to apply, the Employer may assign this Contract.'

Clause 24: Damages for non-completion✳

In clause 24.2.1, delete 'as the Employer may require in writing not later than the date when the Final Statement (or, as the case may be, the Employer's Final Statement) becomes conclusive as to the balance due between the parties by agreement or by the operation of clause 30.5.5'.

Clause 25.4: Relevant Events✳

Clause 25.4.10 shall be deleted.

Add:

'25.4.15 any breach of this Contract by the Employer or any act or omission on the part of the Employer.'

Clause 27: Determination by Employer✳

The proviso to clause 27.1 shall be deleted.

In clause 27.2, the words 'or having an application made under the Insolvency Act 1986 in respect of his company to the court for the appointment of an administrator' shall be deleted, and 'or having an administrator appointed by the court under the Insolvency Act 1986 in respect of his company' shall be substituted.

Clause 30: Payments✳

Clause 30.4.2 shall be deleted, and the following substituted:

'30.4.2   The Retention shall be subject to the following rules:

.2.1   the Employer's interest in the Retention shall not be fiduciary, either as trustee for the Contractor or any other person, or in any other capacity; the relationship of the Employer and the Contractor with regard to the Retention shall be solely that of debtor and unsecured creditor, subject to the terms hereof; and the Employer shall have no obligation to invest the Retention or any part thereof;

.2.2   the Employer shall have no obligation to segregate the Retention or any part thereof in a separate banking account, or in any other manner whatsoever; and shall be entitled to the full beneficial interest in the Retention and every part thereof (and, without limitation, interest thereon and income arising therefrom) unless and until the Retention is paid to the Contractor pursuant to this Contract.'

In clause 30.4.3, delete:

'Notwithstanding the fiduciary interest of the Employer in the Retention as stated in clause 30.4.2.1...'.

In clause 30.8.2, the last word ('either') of the first paragraph shall be deleted.

In clause 30.8.2.1, the last word ('or') shall be deleted.

Clause 30.8.2.2 shall be deleted.

Clause 31: Finance (No.2) Act 1975 – statutory tax deduction scheme

References to 'the Act', 'the Regulations' or any legislation shall be interpreted to include references to any amendment or re-enactment thereof.

Clause 36.4: Provisions relating to clause 36✻

Clause 36.4.8.1 shall be deleted.

Clause 37.5: Provisions relating to clause 37✻

Clause 37.5.8.1 shall be deleted.

Clause 38: Use of price adjustment formulae✻

Clause 38.6.2.1 shall be deleted.

Additional Conditions

The following clauses shall be added:

'40:   Collateral Warranties✻

40.1   The Contractor shall within 7 (seven) working days of the Employer's request so to do:

40.1.1   execute, in favour of any persons who have entered or shall enter into an agreement for the provision of finance in connection with the Works, a Deed in the form annexed as Annexe 'A', or a similar form reasonably required by the Employer, and deliver the same to the Employer, together with a guarantee in the form annexed as Annexe 'C', or a similar form reasonably required by the Employer, from the ultimate parent company of the Contractor, namely [   ] [Limited] **OR** [PLC], in respect of the Contractor's obligations pursuant to such Deed; and

40.1.2   execute, in favour of any persons who have acquired or shall acquire any interest in or over the Works or any part thereof, a Deed in the form annexed as Annexe 'B', or a similar form reasonably required by the Employer, and deliver the same to the

Employer, together with a guarantee in the form annexed as Annexe 'C', or a similar form reasonably required by the Employer, from the ultimate parent company of the Contractor, namely [   ] [Limited] **OR** [PLC], in respect of the Contractor's obligations pursuant to such Deed.

40.2   The Contractor shall, if requested by the Employer, use reasonable endeavours to procure that each sub-contract or supply contract with any sub-contractor or supplier (including, without limitation, any Architect, Engineer, Quantity Surveyor or other person to whom any part of the design or other aspects of the Works may be sub-contracted), shall contain obligations on the relevant sub-contractor or supplier to execute, in favour of the Employer and/or (within 7 (seven) working days of the Employer's request so to do) any persons who have entered or shall enter into an agreement for the provision of finance in connection with the Works and/or in favour of any persons who have acquired or shall acquire any interest in or over the Works or any part thereof, a Deed in the form annexed as Annexe 'D' (or, in the case of any Architect, Engineer, Quantity Surveyor or other consultant, in the form annexed as Annexe 'E') or a similar form reasonably required by the Employer, and deliver the same to the Employer; together in each case (if requested by the Employer) with a guarantee in the form annexed as Annexe 'F', or a similar form reasonably required by the Employer, from the ultimate parent company of the relevant sub-contractor or supplier in respect of the sub-contractor's or supplier's obligations pursuant to such Deed. The Contractor shall enforce such obligations or such modified obligations as are referred to below.

40.3   If, despite the Contractor having used such reasonable endeavours, the sub-contractor or supplier will not accept such obligations, or will only accept them in a modified form, the Contractor shall notify the Employer, who may agree in writing that the relevant sub-contract or supply contract need not contain such obligations, or that the relevant obligations may be in a modified form agreeable to the sub-contractor or supplier.

40.4   Failing such agreement by the Employer, the Contractor shall not enter into a relevant sub-contract or supply contract with that sub-contractor or supplier.

40.5   The above obligations for the provision of Deeds and guarantees in favour of third parties shall continue notwithstanding termination of this Contract, or determination of the Contractor's employment hereunder, in either case for any reason whatsoever, including (without limitation) breach by the Employer. However, any such Deed given after such termination or determination shall be amended by the Employer so as to refer to the fact and date of such termination or determination, to omit any obligation to continue to exercise skill, care or diligence or fulfil obligations pursuant to this Contract after such termination or determination, and to omit any provision enabling a third party to assume the position of employer of the Contractor.

41: **[Performance Bond] OR [Parent Company Guarantee]**

Upon execution of this Contract, the Contractor shall deliver to the Employer a performance bond from [   ] [Limited] **OR** [PLC] in an amount of 10% (ten per cent) of the Contract Sum in the form annexed as Annexe 'G' **OR** parent company guarantee from its ultimate parent company, namely [   ] [Limited] **OR** [PLC], for its performance of this Contract in the form annexed as Annexe 'G'.

42: **Professional Indemnity Insurance✳**

42.1   The Contractor shall maintain professional indemnity insurance covering (inter alia) all its liability hereunder in respect of defects or insufficiency in design upon customary and usual terms and conditions prevailing for the time being in the insurance market, and with reputable insurers lawfully carrying on such insurance business in the United Kingdom, in an amount of not less than   pounds (£   ) for any one occurrence or series of occurrences arising out of any one event for a period beginning now and ending 15 (fifteen) years after the date of practical completion of the Works, provided always that such insurance is available at commercially reasonable rates. The said terms and conditions shall not include any term or condition to the effect that the Contractor must discharge any liability before being entitled to recover from the insurers, or any other term or condition which might adversely affect the rights of any person to recover from the insurers pursuant to the Third Parties (Rights Against Insurers) Act 1930, or any amendment or re-enactment thereof. The Contractor shall not, without the prior approval in writing of the Employer, settle or compromise with the insurers any claim which the Contractor may have against the insurers and which relates to a claim by the Employer against the Contractor, or by any act or omission lose or prejudice the Contractor's right to make or proceed with such a claim against the insurers.

42.2   Any increased or additional premium required by insurers by reason of the Contractor's own claims record or other acts, omissions, matters or things particular to the Contractor shall be deemed to be within commercially reasonable rates.

42.3   The Contractor shall immediately inform the Employer if such insurance ceases to be available at commercially reasonable rates in order that the Contractor and the Employer can discuss means of best protecting the respective positions of the Employer and the Contractor in respect of the Works in the absence of such insurance.

42.4   The Contractor shall fully co-operate with any measures reasonably required by the Employer, including (without limitation) completing any proposals for insurance and associated documents, maintaining such insurance at rates above commercially reasonable rates if the Employer undertakes in writing to reimburse the Contractor in respect of the net cost of such insurance to the Contractor above commercially reasonable rates or, if the Employer effects such insurance at rates at or above commercially reasonable rates, reimbursing the Employer in respect of what the net cost of such insurance to the Employer would have been at commercially reasonable rates.

42.5   As and when reasonably required to do so by the Employer, the Contractor shall produce for inspection documentary evidence (including, if required by the Employer, the originals of the relevant insurance documents) that its professional indemnity insurance is being maintained.

42.6   The above obligations in respect of professional indemnity insurance shall continue notwithstanding termination of the Contract, or determination of the Contractor's employment hereunder, in either case for any reason whatsoever, including (without limitation) breach by the Employer.

43:   Proscribed Materials

43.1   The Contractor warrants:

43.1.1   that he has not used or specified and will not use or specify for use;

43.1.2   that he has exercised and will continue to exercise reasonable skill, care and diligence to see that there are not used;

43.1.3   that he is not aware and has no reason to suspect or believe that there have been or will be used; and

43.1.4   that he will promptly notify the Employer in writing if he becomes aware or has reason to suspect or believe that there have been or will be used;

in or in connection with the Works, any of the materials or substances identified in clause 43.2.

43.2   The said materials or substances are:

43.2.1   high alumina cement in structural elements;

43.2.2 wood wool slabs in permanent formwork to concrete;

43.2.3 calcium chloride in admixtures for use in reinforced concrete;

43.2.4 asbestos products;

43.2.5 naturally occurring aggregates for use in reinforced concrete which do not comply with British Standard 882: 1983 and/or naturally occurring aggregates for use in concrete which do not comply with British Standard 8110: 1985.'

Supplementary Provisions (issued February 1988)∗

Delete S4.4.2, and substitute:

'If the employment of the Named Sub-Contractor is determined the Contractor shall itself complete any balance of the Named Sub-Contract Works left uncompleted at the date of determination. Such completion shall not be treated as if it were work executed in accordance with a Change.'

Delete S4.4.3 and S4.4.4.

Appendix 2: Method of payment – alternatives

In both Alternatives A and B, delete the provisions relating to off-site materials or goods.

APPENDIX 1∗

APPENDIX 2∗

Method of payment – alternatives

APPENDIX 3∗

ANNEXE 'E'✶

[   ]

('the Firm')

-and-

[   ]

(['the Employer'] **OR** ['the Fund'] **OR** ['the Acquirer'])

COLLATERAL WARRANTY BY CONSULTANT
EMPLOYED BY 'DESIGN AND BUILD' CONTRACTOR

in respect of professional services
relating to the design and construction of
[   ]

THIS AGREEMENT is made the   day of   One thousand nine hundred and ninety-

BETWEEN:

(1)

[of] **OR** [whose registered office is at]

('the Firm'); and

(2)

[of] **OR** [whose registered office is at]

(['the Employer'] **OR** ['the Fund'] **OR** ['the Acquirer'], which term shall include its successors and assigns).

WHEREAS:

(A)✱   [The Employer] **OR** [('the Employer')] has entered into a building contract dated   199   ('the Building Contract') with [   ] ('the Contractor') for the design and construction of a project of development briefly described as

at

('the Development').

(B)   By a contract ('the Appointment') dated   199   (a copy of which is annexed and signed for identification purposes by the parties) the Contractor has appointed the Firm as [architects] **OR** [consulting structural engineers] **OR** [consulting building services engineers] **OR** [surveyors] **OR** [project managers] in connection with the Development.

[(C)   The Fund has entered into an agreement with the Employer for the provision of certain finance in connection with the carrying out of the Development.] [The Fund entered into such agreement, and enters into this Agreement, on its own behalf and as agent for a syndicate of banks. Each of the banks which are members of the syndicate from time to time, including banks joining the syndicate after the date of this Agreement, shall be entitled to the benefit of this Agreement in addition to the Fund.]

OR

[(C)   The Acquirer intends to acquire, or has acquired, an interest in the Development.]

NOW in consideration of £1 (one pound) paid by the [Employer] **OR** [Fund] **OR** [Acquirer] to the Firm (receipt of which the Firm hereby acknowledges) **THIS DEED WITNESSETH** as follows:

1.✱   The Firm warrants that it has exercised and will continue to exercise reasonable skill, care and diligence in the performance of its duties to the Contractor under the Appointment.

2.1   Without prejudice to the generality of clause 1, the Firm further warrants:

  2.1.1✱ that it has not specified and will not specify for use;

  2.1.2   that it has exercised and will continue to exercise reasonable skill, care and diligence to see that there are not used;

  2.1.3   that it is not aware and has no reason to suspect or believe that there have been or will be used; and

  2.1.4   that it will promptly notify the [Employer] **OR** [Fund] **OR** [Acquirer] in writing if it becomes aware or has reason to suspect or believe that there have been or will be used;

  in or in connection with the Development, any of the materials or substances identified in clause 2.2.

2.2   The said materials or substances are:

  2.2.1   high alumina cement in structural elements;

  2.2.2   wood wool slabs in permanent formwork to concrete;

  2.2.3   calcium chloride in admixtures for use in reinforced concrete;

  2.2.4   asbestos products;

  2.2.5   naturally occurring aggregates for use in reinforced concrete which do not comply with British Standard 882: 1983 and/or naturally occurring aggregates for use in concrete which do not comply with British Standard 8110: 1985.

3.   The [Employer] **OR** [Fund] **OR** [Acquirer] has no authority to issue any direction or instruction to the Firm in relation to performance of the Firm's duties under the Appointment.

4.   The Firm acknowledges that the Contractor has paid all sums due and owing to the Firm under the Appointment up to the date of this Agreement. The [Employer] **OR** [Fund] **OR** [Acquirer] has no liability to the Firm in respect of sums due under the Appointment.

5.   The copyright in all drawings, reports, specifications, bills of quantities, calculations and other similar documents provided by the Firm in connection with the Development shall remain vested in the Firm (or as may be otherwise provided by the Appointment), but the [Employer] **OR** [Fund] **OR** [Acquirer] and its appointee shall have a licence to copy and use such drawings and other documents, and to reproduce the designs contained in them, for any purpose related to the Development including, but without limitation, the construction, completion, maintenance, letting, promotion, advertisement, reinstatement, repair and/or extension of the Development. The Firm shall, if the [Employer] **OR** [Fund] **OR** [Acquirer] so requests and undertakes in writing to pay the Firm's reasonable copying charges, promptly supply the [Employer] **OR** [Fund] **OR** [Acquirer] with conveniently reproducible copies of all such drawings and other documents.

6.1   The Firm shall maintain professional indemnity insurance covering (inter alia) all liability hereunder upon customary and usual terms and conditions prevailing for the time being in the insurance market, and with reputable insurers lawfully carrying on such insurance business in the United Kingdom, in an amount of not less than   pounds (£   ) for any one occurrence or series of occurrences arising out of any one event for a period beginning now and ending 15 (fifteen) years after the date of practical completion of the Development for the purposes of the Building Contract, provided always that such insurance is available at commercially reasonable rates. The said terms and conditions shall not include any term or condition to the effect that the Firm must discharge any liability before being entitled to recover from the insurers, or any other term or condition which might adversely affect the rights of any person to recover from the insurers pursuant to the Third Parties (Rights Against Insurers) Act 1930, or any amendment or re-enactment thereof. The Firm shall not, without the prior approval in writing of the [Employer] **OR** [Fund] **OR** [Acquirer], settle or compromise with the insurers any claim which the Firm may have against the insurers and which relates to a claim by the [Employer] **OR** [Fund] **OR** [Acquirer] against the Firm, or by any act or omission lose or prejudice the Firm's right to make or proceed with such a claim against the insurers.

6.2   Any increased or additional premium required by insurers by reason of the Firm's own claims record or other acts, omissions, matters or things particular to the Firm shall be deemed to be within commercially reasonable rates.

6.3   The Firm shall immediately inform the [Employer] **OR** [Fund] **OR** [Acquirer] if such insurance ceases to be available at commercially reasonable rates in order that the Firm and the [Employer] **OR** [Fund] **OR** [Acquirer] can discuss means of best protecting the respective positions of the [Employer] **OR** [Fund] **OR** [Acquirer] and the Firm in respect of the Development in the absence of such insurance.

6.4   The Firm shall fully co-operate with any measures reasonably required by the [Employer] **OR** [Fund] **OR** [Acquirer], including (without limitation) completing any proposals for insurance and associated documents, maintaining such insurance at rates above commercially reasonable rates if the [Employer] **OR** [Fund] **OR** [Acquirer] undertakes in writing to reimburse the Firm in respect of the net cost of such insurance to the Firm above commercially reasonable rates or, if the [Employer] **OR** [Fund] **OR** [Acquirer] effects such insurance at rates at or above commercially reasonable rates, reimbursing the [Employer] **OR** [Fund] **OR** [Acquirer] in respect of what the net cost of such insurance to the [Employer] **OR** [Fund] **OR** [Acquirer] would have been at commercially reasonable rates.

6.5   As and when it is reasonably requested to do so by the [Employer] **OR** [Fund] **OR** [Acquirer] the Firm shall produce for inspection documentary evidence (including, if required by the [Employer] **OR** [Fund] **OR** [Acquirer], the original of the relevant insurance documents) that its professional indemnity insurance is being maintained.

7.   This Agreement may be assigned by the [Employer] **OR** [Fund] **OR** [Acquirer] and its successors and assigns without the consent of the Firm being required.

8.   Any notice to be given by the Firm hereunder shall be deemed to be duly given if it is delivered by hand at or sent by registered post or recorded delivery to the above-mentioned address of the [Employer] **OR** [Fund] **OR** [Acquirer] or to the principal business address of the [Employer] **OR** [Fund] **OR** [Acquirer] for the time being, and any notice to be given by the [Employer] **OR** [Fund] **OR** [Acquirer] hereunder shall be deemed to be duly given if it is addressed to the Firm and delivered by hand at or sent by registered post or recorded delivery to the above-mentioned address of the Firm or to the principal business address of the Firm for the time being and, in the case of any such notices, the same shall if sent by registered post or recorded delivery be deemed to have been received forty-eight hours after being posted.

NOTE: The following clause will only be required in the case of Collateral Warranties in favour of the Employer or Fund.

9.   The Firm shall within 7 (seven) working days of the [Employer's] **OR** [Fund's] request to do so, execute, in favour of any persons who have entered or shall enter into an agreement for the provision of finance in connection with the Development and/or in favour of any persons who have acquired or shall acquire any interest in or over the Development or any part thereof, a Deed in the form of this Deed, excluding this clause, or a similar form reasonably required by the [Employer] **OR** [Fund], and deliver the same duly executed to the [Employer] **OR** [Fund]; together in each case (if requested by the [Employer] **OR** [Fund]) with a guarantee (in form and substance reasonably required by the [Employer] **OR** [Fund]) from the ultimate parent company of the Firm in respect of the Firm's obligations pursuant to such Deed.

IN WITNESS whereof the Firm has executed this Deed on the date first stated above.

[ANNEXE 'G'*

DATED **199**

[   ]

('the Contractor')

-and-

[   ]

('the Surety')

-and-

[   ]

('the Employer')

PERFORMANCE BOND
in respect of the design and
construction of

[   ]

THIS BOND is made the   day of   One thousand nine hundred and ninety-

BETWEEN:

(1)

        [of] **OR** [whose registered office is at]

        ('the Contractor');

(2)

        [of] **OR** [whose registered office is at]

        ('the Surety'); and

(3)

        [of] **OR** [whose registered office is at]

        ('the Employer', which term shall include its successors and assigns).

WHEREAS by an Agreement ('the Contract') dated   199   and made between the Employer of the one part and the Contractor of the other part, the Contractor undertook the design and construction of certain Works in accordance with the terms and conditions of the Contract.

NOW THIS DEED WITNESSETH as follows:

1   Bond

By this Bond the Contractor and the Surety, their successors and assigns, are jointly and severally held and bound to the Employer for payment to the Employer of the sum of   pounds (£   ).

2   Conditions

The conditions of this Bond are that if:

2.1   the Contractor duly discharges all the Contractor's obligations under or pursuant to the Contract; or

2.2   in the event of the Contractor's default in the discharge of any such obligations, the Surety shall pay to the Employer the loss and damage thereby caused to the Employer, up to the amount of this Bond; or

2.3   pursuant to clause 30.8.1 of the Contract, the Final Account and Final Statement have been agreed or have become conclusive as to the balance due between the Employer and the Contractor in accordance with clause 30.5.5 of the Contract, or the Employer's Final Account and Employer's Final Statement have become conclusive as to the balance due between the Employer and the Contractor in accordance with clause 30.5.8 of the Contract, and (without any undetermined or unresolved exception as provided in clauses 30.8.2 or 30.8.3 of the Contract, and save in respect of fraud) would have effect in any proceedings arising out of, or in connection with, the Contract (whether by arbitration under Article 5 thereof or otherwise) as conclusive evidence of the matters described in clauses 30.8.1.1, 30.8.1.2 and 30.8.1.3 of the Contract, and any balance due from the Contractor to the Employer (having been adjusted, if necessary, in accordance with clauses 30.8.2 and/or 30.8.3 of the Contract) has been paid; and any amount due from the Contractor to the Employer pursuant to any award or judgment in, or settlement of, any arbitration or other proceedings commenced in respect of the Contract before or within 28 (twenty-eight) days after the Final Account and Final Statement, or the Employer's Final Account and Employer's Final Statement, as the case may be, would (if not disputed) otherwise have become conclusive by the operation of clause 30.5.5 or 30.5.8 of the Contract, has been paid;

this Bond shall thereby be discharged, but otherwise shall remain in force.

3   Alterations

No alterations in the Contract, or in the Works, and no extension of time, forbearance or forgiveness, nor any act, matter or thing whatsoever except fulfilment of one of the above conditions or an express release by Deed by the Employer, shall in any way release the Surety from any liability under this Bond.

[4   Reduction on Practical Completion

Provided that upon Practical Completion of the Works under the Contract, the amount of this Bond shall reduce by one-half.]

IN WITNESS whereof the Contractor and the Surety have executed this Deed on the date first stated above.]

OR

[ANNEXE 'G'✳

DATED _____ **199**

[   ]

('the Guarantor')

-and-

[   ]

('the Employer')

GUARANTEE

in respect of the design and construction of
[   ]

THIS AGREEMENT is made the   day of   One thousand nine hundred and ninety-

BETWEEN:

(1)

[of] **OR** [whose registered office is at]

('the Guarantor'); and

(2)

[of] **OR** [whose registered office is at]

('the Employer', which term shall include its successors and assigns).

WHEREAS by an Agreement ('the Contract') dated   199   and made between the Employer of the one part and [   ] ('the Contractor') of the other part, the Contractor undertook the design and construction of certain Works in accordance with the terms and conditions of the Contract.

NOW THIS DEED WITNESSETH that if the Contractor defaults in the discharge of any of the Contractor's obligations under or pursuant to the Contract, the Guarantor will indemnify the Employer against all loss and damage thereby caused to the Employer, and no alterations in the Contract, or in the Works, and no extension of time, forbearance or forgiveness, nor any act, matter or thing whatsoever except an express release by Deed by the Employer, shall in any way release the Guarantor from any liability hereunder.

IN WITNESS whereof the Guarantor has executed this Deed on the date first stated above.]

COMMENTARY

Clause 1.3: Definitions

See paragraph 6, Chapter 2.

Clause 2.2: Employers Requirements, etc. – relation to Articles, Conditions and Appendices

See paragraph 7, Chapter 2.

Clause 2.5.1: Contractor's design warranty

This very important sub-clause is left unamended, but should be explained to the Employer. See paragraph 11 of Chapter 1 regarding the vital subject of design responsibility and possible warranties of fitness for purpose.

Clause 18.1.1 and 18.1.2: Assignment

See paragraphs 8 and 9, Chapter 2.

Clause 24: Damages for non-completion

See paragraph 10, Chapter 2.

Clause 25.4: Relevant Events

See paragraphs 11, 12 and 13, Chapter 2.

Clause 27: Determination by Employer

See paragraphs 15 and 16, Chapter 2.

Clause 30: Payments

See paragraphs 22, 23 and 24, Chapter 2.

Clause 36.4: Provisions relating to clause 36
Clause 37.5: Provisions relating to clause 37
Clause 38: Use of price adjustment formulae

See paragraph 25, Chapter 2.

Clause 40: Collateral Warranties

For Annexes 'A', 'B', 'C', 'D' and 'F', see Forms 19, 20, 28, 21 and 29.

Clause 42: Professional Indemnity Insurance

See commentary upon clause 9 of Form 6.

Supplementary Provisions S4.4.2, S4.4.3 and S4.4.4

The amendments (which the Contractor will no doubt resist) cast the financial consequences of the determination of the employment of a Named Sub-Contractor upon the Contractor.

Appendix 1

The published Appendix should be completed and attached. With regard to the entry re: clause 18.1.2, delete 'of benefits after Practical Completion'. With regard to the periods of delay under clauses 28 and 28A, the

Employer may wish to insert longer periods (e.g. six months) than those suggested in the footnotes to the JCT form. It is suggested that the appointor of arbitrators should be the President or a Vice-President of the Chartered Institute of Arbitrators, in view of that Institute's training of its arbitrators.

Appendix 2: Method of payment – alternatives

The published Appendix should be completed and attached.

Appendix 3

The published Appendix should be completed and attached.

Annexe 'E': Collateral Warranty by Consultant employed by 'Design and Build' Contractor

Annexe 'E' is very much along the lines of Form 21, but 'Firm' and 'Appointment' have been substituted for 'Sub-Contractor' and 'Sub-Contract'.

Recital (A)

Refers to 'design and construction'.

Clause 1

Unlike clause 1 of Form 21, this clause is restricted to a duty of care, which will be the normal measure of the Firm's duty under the Appointment.

Clause 2.1.1

Unlike clause 2.1.1 of Form 21, this sub-clause refers only to specification for use.

Annexe 'G': Performance Bond or Parent Company Guarantee

Cf. Forms 30 and 31.

FORM 16: MAIN CONTRACT INCORPORATING JCT MANAGEMENT CONTRACT 1987, WITH OPTIONS FOR PHASED COMPLETION, INCLUDING PERFORMANCE BOND AND PARENT COMPANY GUARANTEE

DATED _____ **199**

[   ]

('the Employer')

-and-

[   ]

('the Management Contractor')

MAIN CONTRACT

for the construction of

[   ]

✱ An asterisk indicates that there is a relevant note in the commentary upon this Form.

THIS AGREEMENT made the   day of   One thousand nine hundred and ninety-

BETWEEN:

(1)

[of] **OR** [whose registered office is at]

('the Employer'); and

(2)

[of] **OR** [whose registered office is at]

('the Management Contractor');

INCORPORATES the Joint Contracts Tribunal Standard Form of Management Contract, 1987 Edition, as printed in   199   , incorporating Amendments 1 (July 1988) and 2 (April 1989), [as modified by the JCT Phased Completion Supplement issued in December 1987 and] as amended by the [Sixth] **OR** [Seventh] Schedule.

IN WITNESS whereof the parties have executed this Deed in duplicate on the date first stated above.

APPENDIX: PART 1✻

APPENDIX: PART 2✻

FIRST SCHEDULE

Description of the Project

SECOND SCHEDULE

Definition of Prime Cost payable to the Management Contractor

The Second Schedule applies [unaltered.] **OR** [altered as follows:]

[The lists referred to in Part 1, paragraph 3(ii) and Part 3A, paragraph 1.1 of the Second Schedule are attached.]

THIRD SCHEDULE

Services Provided or to be Provided by the Management Contractor

The Third Schedule applies [unaltered.] **OR** [altered as follows:]

FOURTH SCHEDULE

List of Project Drawings

FIFTH SCHEDULE✳

Site Facilities and Services to be Provided by the Management Contractor

---

[SIXTH SCHEDULE✳

Phased Completion]

[SIXTH] [SEVENTH] SCHEDULE✻

Articles of Agreement

[Article 3A: Architect✻

The name and address of the Architect are

[of] **OR** [whose registered office is at]

Delete 'not being a person to whom the Management Contractor no later than 7 days after such nomination shall object for reasons considered to be sufficient by an Arbitrator appointed in accordance with Section 9'.]

OR

[Article 3B: Contract Administrator✻

The name and address of the Contract Administrator are

[of] **OR** [whose registered office is at]

Delete 'not being a person to whom the Management Contractor no later than 7 days after such nomination shall object for reasons considered to be sufficient by an Arbitrator appointed in accordance with Section 9'.]

Article 4: Quantity Surveyor✻

The name and address of the Quantity Surveyor are

[of] **OR** [whose registered office is at]

Delete 'not being a person to whom the Management Contractor no later than 7 days after such nomination shall object for reasons considered to be sufficient by an Arbitrator appointed in accordance with Section 9'.

Article 5: Professional Team

The following names and addresses shall be inserted:

Amendments to Conditions

Clause 1.3: Definitions

The definition of 'Conditions' shall include any additional clauses or provisions hereby added, and the Conditions as hereby amended and added to.

Add 'his successors and assigns' to the definition of 'Employer'.✶

The definition of 'Joint Names Policy' shall be deleted, and the following substituted:

'Joint Names Policy: a policy of insurance which includes the Management Contractor and the Employer and such other persons as the Employer may reasonably require (including, without limitation, any persons who have entered or shall enter into an agreement for the provision of finance in connection with the Project, and any persons who have acquired or shall acquire any interest in or over the Project or any part thereof) as the insured.'

Clause 1.14: Effect of Final Certificate✶

In clause 1.14.2, the last word ('either') of the first paragraph shall be deleted.

In clause 1.14.2.1, the last word ('or') shall be deleted.

Clause 1.14.2.2 shall be deleted.

The concluding words of clause 1.14.2 ('whichever is the earlier') shall be deleted.

Clause 2.10: Liquidated and ascertained damages✶

Delete 'as the Employer may require in writing not later than the date of the Final Certificate'.

Clause 2.13: Project Extension Items✱

In clause 2.13.2, the opening words shall be deleted, and the following substituted:

> '.2 any Relevant Event, except the Relevant Events referred to in clauses 2.10.7.1 or 2.10.10 of the Works Contract Conditions, ...'

Clause 3.6: Acceleration, etc.✱

Clause 3.6.4 shall be deleted.

In clause 3.6.5, the words '(or after receipt of a Preliminary Instruction re-issued under clause 3.6.4)' shall be deleted.

Clause 3.6.6 shall be deleted, and the following substituted:

> 'On receipt of the information given to the [Architect] **OR** [Contract Administrator] under clause 3.6.5, or if such information is not given as soon as reasonably practicable after receipt of the Preliminary Instruction by the Management Contractor, and whether or not the Employer agrees to pay the amounts referred to in clause 3.6.5.1 or to accept the Completion Date stated by the Management Contractor pursuant to clause 3.6.5.2, the Employer may cause the [Architect] **OR** [Contract Administrator] to issue an Instruction:
>
> – confirming or stating the details of the acceleration or alteration of sequence or timing required, including the change or changes to any Works Contract period or periods for completion of the Works Contract Works (whether or not such change or changes are as stated by Works Contractors in response to the Management Contractor under clause 3.4.6.2 of the Works Contract Conditions); and
> – fixing the Completion Date [of the Phase].
>
> Such an Instruction shall not, unless it expressly states to the contrary, be evidence of agreement by the Employer to pay the amounts referred to in clause 3.6.5.1. In the absence of such agreement, the cost to the Employer of compliance by Works Contractors with such an Instruction shall be ascertained in accordance with all the relevant Works Contract Conditions.'

Clauses 3.19 and 3.20: Assignment✱

Delete clause 3.19 and the first sentence of clause 3.20, and substitute:

'3.19   The Management Contractor shall not, without the written consent of the Employer, assign this Contract.

3.20   Where clause 3.20 is stated in the Appendix to apply, the Employer may assign this Contract.'

[Clause 3.21: Breach of Contract by Works Contractor, etc.✻]

Clauses 4.3 and 4.8: Retention✻

In clause 4.3.2 delete:

'Notwithstanding the fiduciary interest of the Employer in the Retention as stated in clause 4.8.1...'.

Clause 4.8.1 shall be deleted, and the following substituted:

'4.8.1   the Employer's interest in the Retention shall not be fiduciary, either as trustee for the Management Contractor or any Works Contractor or any other person, or in any other capacity; the relationship of the Employer and the Management Contractor with regard to the Retention shall be solely that of debtor and unsecured creditor, subject to the terms hereof; and the Employer shall have no obligation to invest the Retention or any part thereof;'

Clause 4.8.3 shall be deleted, and the following substituted:

'4.8.3   the Employer shall have no obligation to segregate the Retention or any part thereof in a separate banking account, or in any other manner whatsoever; and shall be entitled to the full beneficial interest in the Retention and every part thereof (and, without limitation, interest thereon and income arising therefrom) unless and until the Retention is paid to the Management Contractor pursuant to this Contract.'

[Clause 4.10: Any adjustment of Construction Period Management Fee✻

Delete, and substitute:]

Clause 7.1: Default of Management Contractor✳

The proviso to clause 7.1 shall be deleted.

Clause 7.2: Management Contractor becoming insolvent✳

Add 'becoming bankrupt' after 'Management Contractor', where it first appears.

Delete 'or having an application made under the Insolvency Act 1986 in respect of his company to the court for the appointment of an administrator', and substitute 'or having an administrator appointed by the court under the Insolvency Act 1986 in respect of his company'.

Add 'his trustee in bankruptcy' after 'Management Contractor', where it last appears.

Clause 7.4: Determination of employment of Management Contractor, etc.✳

In clause 7.4.2.1, delete 'Management Contractor' in the first line and substitute:

'bankruptcy of the Management Contractor or of him'.

Clause 7.5: Default of Employer, etc.✳

In clause 7.5.4, add 'becomes bankrupt' after 'Employer', where it first appears.

Delete 'or has an application made under the Insolvency Act 1986 in respect of his company to the court for the appointment of an administrator' and substitute 'or has an administrator appointed by the court under the Insolvency Act 1986 in respect of his company'.

Clause 7.6: Determination of employment of Management Contractor under clause 7.5, etc.✳

In clause 7.6.2.4, the last word ('and') shall be deleted.

Clause 7.6.2.5 shall be deleted.

Clause 8.2: Selection of Works Contractors, etc.✱

Clause 8.2.1 shall be deleted, and the following substituted:

'The Works Contractors to carry out the items of work so identified shall be selected by the [Architect] **OR** [Contract Administrator] after consultation with, or upon the recommendation of, the Management Contractor, and that selection shall be confirmed in an Instruction. Provided that, save where the Employer or the [Architect] **OR** [Contract Administrator] on his behalf instructs otherwise, the Management Contractor shall only employ any persons as Works Contractors who will:

.1   enter into a contract on the current standard Form (amended only as may be required by the Employer or by the [Architect] **OR** [Contract Administrator] on his behalf) of Works Contract (Works Contract/1 and Works Contract/2) issued by the Joint Contracts Tribunal with the Management Contractor and execute that contract as a Deed; and

.2   if so required (as recorded in Works Contract/1) enter into an Employer/Works Contractor Agreement (Works Contract/3) (amended only as may be required by the Employer or by the [Architect] **OR** [Contract Administrator] on his behalf) with the Employer and execute that Agreement as a Deed'.

Additional Conditions

The following shall be added:

'Section 10: Collateral Warranties✱

10.1   The Management Contractor shall within 7 (seven) working days of the Employer's request so to do:

10.1.1   execute, in favour of any persons who have entered or shall enter into an agreement for the provision of finance in connection with the Project, a Deed in the form annexed as Annexe 'A', or a similar form reasonably required by the Employer, and deliver the same to the Employer, together with a guarantee in the form annexed as Annexe 'C', or a similar form reasonably required by the Employer, from the ultimate parent company of the Contractor, namely [   ] [Limited] **OR** [PLC], in respect of the Contractor's obligations pursuant to such Deed; and

10.1.2 execute, in favour of any persons who have acquired or shall acquire any interest in or over the Project or any part thereof, a Deed in the form annexed as Annexe 'B', or a similar form reasonably required by the Employer, and deliver the same to the Employer, together with a guarantee in the form annexed as Annexe 'C', or a similar form reasonably required by the Employer, from the ultimate parent company of the Contractor, namely [   ] [Limited] **OR** [PLC], in respect of the Contractor's obligations pursuant to such Deed.

10.2 The Management Contractor shall (without request) use reasonable endeavours to procure that each sub-contract or supply contract with each Works Contractor, and any other sub-contractor or supplier (if requested by the Employer), shall contain obligations on the relevant sub-contractor or supplier to execute, in favour of the Employer and/or (within 7 (seven) working days of the Employer's request so to do) any persons who have entered or shall enter into an agreement for the provision of finance in connection with the Project and/or in favour of any persons who have acquired or shall acquire any interest in or over the Project or any part thereof, a Deed in the form annexed as Annexe 'D', or a similar form reasonably required by the Employer, and deliver the same to the Employer; together in each case (if requested by the Employer) with a guarantee in the form annexed as Annexe 'E', or a similar form reasonably required by the Employer, from the ultimate parent company of the relevant sub-contractor or supplier in respect of the sub-contractor's or supplier's obligations pursuant to such Deed. The Management Contractor shall enforce such obligations, or such modified obligations as are referred to below.

10.3 If, despite the Management Contractor having used such reasonable endeavours, the sub-contractor or supplier will not accept such obligations, or will only accept them in a modified form, the Management Contractor shall notify the Employer, who may agree in writing that the relevant sub-contract or supply contract need not contain such obligations, or that the relevant obligations may be in a modified form agreeable to the sub-contractor or supplier.

10.4 Failing such agreement by the Employer in the case of a proposed Works Contractor, the Management Contractor shall not be obliged to enter into a sub-contract or supply contract with that Works Contractor.

10.5 Failing such agreement by the Employer in the case of any other sub-contractor or supplier, the Management Contractor shall not enter into a relevant sub-contract or supply contract with that sub-contractor or supplier.

10.6 The above obligations for the provision of Deeds and guarantees in favour of third parties shall continue notwithstanding termination of this Contract, or determination of the Management Contractor's employment hereunder, in either case for any reason whatsoever, including (without limitation) breach by the Employer. However, any such Deed given after such termination or determination shall be amended by the Employer so as to refer to the fact and date of such termination or determination, to omit any obligation to continue to exercise skill, care or diligence or fulfil obligations pursuant to this Contract after such termination or determination, and to omit any provision enabling a third party to assume the position of employer of the Management Contractor.

Section 11: [Performance Bond] OR [Parent Company Guarantee]✱

Upon execution of this Contract, the Management Contractor shall deliver to the Employer a performance bond from [   ] [Limited] **OR** [PLC] in the amount of £   (pounds) in the form annexed as Annexe 'F' **OR** parent company guarantee from its ultimate parent company, namely [   ] [Limited] **OR** [PLC], for its performance of this Contract in the form annexed as Annexe 'F'.

Section 12: Proscribed Materials

12.1   The Management Contractor warrants:

12.1.1   that he has not used or specified and will not use or specify for use;

12.1.2   that he has exercised and will continue to exercise reasonable skill, care and diligence to see that there are not used;

12.1.3   that he is not aware and has no reason to suspect or believe that there have been or will be used; and

12.1.4   that he will promptly notify the Employer in writing if he becomes aware or has reason to suspect or believe that there have been or will be used;

in or in connection with the Project or Works, any of the materials or substances identified in Clause 12.2.

12.2   The said materials or substances are:

12.2.1   high alumina cement in structural elements;

12.2.2   wood wool slabs in permanent formwork to concrete;

12.2.3   calcium chloride in admixtures for use in reinforced concrete;

12.2.4   asbestos products;

12.2.5   naturally occurring aggregates for use in reinforced concrete which do not comply with British Standard 882: 1983 and/or naturally occurring aggregates for use in concrete which do not comply with British Standard 8110: 1985.'

[ANNEXE 'F'*

DATED   **199**

[   ]

('the Contractor')

-and-

[   ]

('the Surety')

-and-

[   ]

('the Employer')

PERFORMANCE BOND

in respect of the construction of
[   ]

THIS BOND is made the   day of   One thousand nine hundred and ninety-

BETWEEN:

(1)

[of] **OR** [whose registered office is at]

('the Contractor');

(2)

[of] **OR** [whose registered office is at]

('the Surety'); and

(3)

[of] **OR** [whose registered office is at]

('the Employer', which term shall include its successors and assigns).

WHEREAS by an Agreement ('the Contract') dated   199   and made between the Employer of the one part and the Contractor of the other part, the Contractor undertook the construction of a certain Project and Works in accordance with the terms and conditions of the Contract.

NOW THIS DEED WITNESSETH as follows:

1   Bond

By this Bond the Contractor and the Surety, their successors and assigns, are jointly and severally held and bound to the Employer for payment to the Employer of the sum of   pounds (£   ).

2   Conditions

The conditions of this Bond are that if:

2.1   the Contractor duly discharges all the Contractor's obligations under or pursuant to the Contract; or

2.2   in the event of the Contractor's default in the discharge of any such obligations, the Surety shall pay to the Employer the loss and damage thereby caused to the Employer, up to the amount of this Bond; or

2.3   pursuant to clause 1.14 of the Contract, the Final Certificate (without any undetermined or unresolved exception as provided in clauses 1.14.2 or 1.14.3 of the Contract, and save in respect of fraud) would have effect in any proceedings arising out of, or in connection with the Contract (whether by arbitration under Section 9 thereof or otherwise) as conclusive evidence of the matters described in clauses 1.14.1.1, 1.14.1.2, 1.14.1.3 and 1.14.1.4 of the Contract, and any balance due from the Contractor to the Employer pursuant to the Final Certificate (having been adjusted, if necessary, in accordance with clauses 1.14.2 and/or 1.14.3 of the Contract) has been paid; and any amount due from the Contractor to the Employer pursuant to any award or judgment in, or settlement of, any arbitration or other proceedings commenced in respect of the Contract before or within 28 (twenty-eight) days after the said Final Certificate was issued, has been paid;

this Bond shall thereby be discharged, but otherwise shall remain in force.

3   Alterations

No alterations in the Contract, or in the Project or Works, and no extension of time, forbearance or forgiveness, nor any act, matter or thing whatsoever except fulfilment of one of the above conditions or an express release by Deed by the Employer, shall in any way release the Surety from any liability under this Bond.

[4   Reduction on Practical Completion]

[Provided that upon certification of Practical Completion of the Project under the Contract, the amount of this Bond shall reduce by one-half] **OR** [Provided that upon certification of Practical Completion of each Phase of the Project under the Contract, the amount of this Bond shall reduce by the following respective amounts:

£

Phase 1

Phase 2   ]

IN WITNESS whereof the Contractor and the Surety have executed this Deed on the date first stated above.]

OR

[ANNEXE 'F'✱

DATED **199**

[   ]

('the Guarantor')

-and-

[   ]

('the Employer')

GUARANTEE

in respect of the construction of
[   ]

THIS AGREEMENT is made the   day of   One thousand nine hundred and ninety-

BETWEEN:

(1)

      [of] **OR** [whose registered office is at]

      ('the Guarantor'); and

(2)

      [of] **OR** [whose registered office is at]

      ('the Employer', which term shall include its successors and assigns).

WHEREAS by an Agreement ('the Contract') dated   199   and made between the Employer of the one part and [   ] ('the Contractor') of the other part, the Contractor undertook the construction of a certain Project and Works in accordance with the terms and conditions of the Contract.

NOW THIS DEED WITNESSETH that if the Contractor defaults in the discharge of any of the Contractor's obligations under or pursuant to the Contract, the Guarantor will indemnify the Employer against all loss and damage thereby caused to the Employer, and no alterations in the Contract, or in the Project or Works, and no extension of time, forbearance or forgiveness, nor any act, matter or thing whatsoever except an express release by Deed by the Employer, shall in any way release the Guarantor from any liability hereunder.

IN WITNESS whereof the Guarantor has executed this Deed on the date first stated above.]

COMMENTARY

Appendix: Part 1

The published Appendix: Part 1 should be completed and attached. With regard to the entry re: clause 3.20, delete 'of benefits after Practical Completion' or, if the JCT Phased Completion Completion Supplement is being used, the equivalent words in the Appendix: Part 1 to the Supplement. With regard to the periods of delay under clauses 7.5 and 7.7, the Employer may wish to insert longer periods (e.g. six months) than those suggested in the footnotes to the JCT forms. It is suggested that the appointor of arbitrators should be the President or Vice-President of the Chartered Institute of Arbitrators, in view of that Institute's training of its arbitrators.

Appendix: Part 2

The published Appendix: Part 2 should be completed and attached.

Fifth Schedule

JCT Practice Note MC/1 gives a model for the completion of the Fifth Schedule.

Sixth Schedule

Part of the JCT Phased Completion Supplement, if used.

[Sixth] [Seventh] Schedule

This will be the Sixth Schedule if the JCT Phased Completion Supplement is not used, but the Seventh Schedule if it is used.

Article 3A: Architect
Article 3B: Contract Administrator
Article 4: Quantity Surveyor

See paragraph 5, Chapter 2.

Clause 1.3: Definitions

See paragraph 6, Chapter 2.

Clause 1.14: Effect of Final Certificate

See paragraph 24, Chapter 2.

Clause 2.10: Liquidated and ascertained damages

See paragraph 10, Chapter 2.

Clause 2.13: Project Extension Items

See paragraph 12, Chapter 2.

Clause 3.6: Acceleration, etc.

As published, clause 3.6 is merely an 'agreement to agree'. See the opening words of clause 3.6.6.

Clauses 3.19 and 3.20: Assignment

See paragraphs 8 and 9, Chapter 2.

Clause 3.21: Breach of Contract by Works Contractor, etc.

This very important clause is left unamended, but should be explained to the Employer. If it is amended in any particular case, there are cross-references to it in clauses 1.7, 2.10, 3.11 and 3.12. See also Chapter 5.

Clauses 4.3 and 4.8: Retention

See paragraph 22, Chapter 2.

Clause 4.10: Any adjustment of Construction Period Management Fee

This matter would usually be negotiated. Clause 4.10, as published by the JCT, increases the Construction Period Management Fee if Prime Cost exceeds the Contract Cost Plan Total by more than 5%, and vice versa. For example, if the cost overrun is 8%, the Construction Period Management Fee is increased by 3%. In order to give the Management Contractor an incentive to keep Prime Cost down, sometimes his fee is fixed, or a guaranteed maximum price is established.

Clause 7.1: Default of Management Contractor

See paragraph 16, Chapter 2.

Clause 7.2: Management Contractor becoming insolvent

This clause, as published, is only suitable for corporate, not individual, insolvency. See footnote n.1 to p.43 of the JCT form. The published clause also provides for automatic determination upon an application being made for the appointment of an administrator, rather than his appointment. The clause has been amended to deal with both points. See also paragraph 15, Chapter 2.

Clause 7.4: Determination of employment of Management Contractor, etc.

This now provides for individual, as well as corporate, insolvency.

Clause 7.5: Default of Employer, etc.

See commentary upon clause 7.2, above.

Clause 7.6: Determination of employment of Management Contractor under clause 7.5, etc.

Under clause 7.6.2.3, the Management Contractor receives appropriate fees. Clause 7.6.2.5 would enable him to recover loss of profit on fees not earned.

Clause 8.2: Selection of Works Contractors, etc.

The amended clause enables Works Contractors to be selected by the Architect or the Contract Administrator, rather than leaving their selection to an 'agreement to agree'. The Employer is also enabled to require amendments to the standard JCT Works Contract documentation. Certain amendments to the Works Contract Conditions (Works Contract/2) follow from amendments to the JCT Management Contract.

Section 10: Collateral Warranties

For Annexes 'A', 'B', 'C', 'D' and 'E', see Forms 19, 20, 28, 21 and 29.

Section 11: Performance Bond or Parent Company Guarantee

As there is no stated Contract Sum under a management contract, a percentage for a performance bond cannot be specified.

Annexe 'F': Performance Bond or Parent Company Guarantee

Cf. Forms 30 and 31.

FORM 17: MAIN CONTRACT INCORPORATING JCT INTERMEDIATE FORM OF BUILDING CONTRACT 1984, WITH OPTIONS FOR SECTIONAL COMPLETION, INCLUDING PERFORMANCE BOND

DATED _____ **199**

[   ]

('the Employer')

-and-

[   ]

('the Contractor')

MAIN CONTRACT

for the construction of

[   ]

✱ An asterisk indicates that there is a relevant note in the commentary upon this Form.

THIS AGREEMENT made the　　　　　　　　day of　　　　　　　　One thousand nine hundred and ninety-

BETWEEN:

(1)

[of] **OR** [whose registered office is at]

('the Employer', which term shall include its successors and assigns); and

(2)

[of] **OR** [whose registered office is at]

('the Contractor');

INCORPORATES the Joint Contracts Tribunal Intermediate Form of Building Contract, 1984 Edition, as printed in　　　　　　　　199　, incorporating Amendments 1 (November 1986), 2 (September 1987), 3 (July 1988), 4 (July 1988), 5 (April 1989) and 6 (July 1991), [as modified by the IFC/SCS Sectional Completion Supplement, as printed in 199　,] and as amended by the Schedule.

IN WITNESS whereof the parties have executed this Deed in duplicate on the date first stated above.

SCHEDULE

Recitals

First recital

The work is the construction of

at

The documents showing and describing the work are:

− the Contract Drawings numbered:

[–   the Specification.]

OR

[–   the Schedules of Work.]

OR

[–   Bills of Quantities.]

Second recital

[Alternative A shall apply. The Contractor has priced the [Specification] **OR** [Schedules of Work] **OR** [Bills of Quantities].]

OR

[Alternative B shall apply. The Contractor has supplied to the Employer a [Contract Sum Analysis] **OR** [Schedule of Rates].]

Articles of Agreement

Article 2: Contract Sum

The sum to be inserted is   pounds (£   ).

Article 3: [The Architect] OR [Contract Administrator]*

The name and address of the [Architect] **OR** [Contract Administrator] are

[of] **OR** [whose registered office is at]

Delete 'not being a person to whom the Contractor shall object for reasons considered to be sufficient by an Arbitrator appointed in accordance with Article 5'.

Article 4: Quantity Surveyor*

The name and address of the Quantity Surveyor are

[of] **OR** [whose registered office is at]

Delete 'not being a person to whom the Contractor shall object for reasons considered to be sufficient by an Arbitrator appointed in accordance with Article 5.'

Amendments to Conditions

Clause 1.3: Priority of Contract Documents*

Delete.

Clause 2.4: Events referred to in 2.3*

Clause 2.4.10 shall be deleted.

Clause 2.4.11 shall be deleted.

Add to clause 2.4:

'2.4.16   any breach of this Contract by the Employer or any act or omission on the part of the Employer, the [Architect] **OR** [Contract Administrator] or the Quantity Surveyor.'

Clause 2.7: Liquidated damages for non-completion*

In clause 2.7, delete 'as the Employer may require in writing not later than the date of the final certificate for payment'.

Clause 3.1: Assignment*

Delete, and substitute:

'3.1   The Contractor shall not, without the written consent of the Employer, assign this Contract. The Employer may assign this Contract.'

[Clause 3.3: Named persons as sub-contractors✽

Clause 3.3.4, except the final paragraph thereof, shall be deleted, and the following substituted:

'3.3.4   The Contractor shall be entitled to no right, relief or remedy whatsoever by reason of the determination of the employment of any named person, or by reason of the giving of any instructions pursuant to clause 3.3.3.'

Clause 3.3.5 shall be deleted.

Clause 3.3.6 shall be deleted.

In clause 3.3.7, the middle part of the first sentence shall be amended to read as follows:

'the Contractor shall be responsible to the Employer under this Contract for anything to which the above terms relate and, through the Contractor, the person so named shall also be so responsible;']

Clause 4.2: Interim Payment✽

Clause 4.2.1(c) shall be deleted.

Clause 4.4: Interest in percentage withheld✽

Delete, and substitute:

'4.4   The Employer's interest in the percentage of the total value not included in the amounts of the interim payments to be certified under clauses 4.2 and 4.3 shall not be fiduciary, in any capacity; the relationship of the Employer and the Contractor with regard to the same shall be solely that of debtor and unsecured creditor, subject to the terms hereof; the Employer shall have no obligation to invest the same or any part thereof; the Employer shall have no obligation to segregate the same or any part thereof in a separate banking account, or in any other manner whatsoever; and shall be entitled to the full beneficial interest in the same and every part thereof (and, without limitation, interest thereon and income arising therefrom) unless and until the same is paid to the Contractor pursuant to this Contract.'

Clause 7.1: Determination by Employer✽

The proviso to clause 7.1 shall be deleted.

Clause 7.2: Contractor becoming bankrupt, etc.✽

In clause 7.2, the words 'or has an application made under the Insolvency Act 1986 in respect of his company to the court for the appointment of an administrator' shall be deleted, and 'or has an administrator appointed by the court under the Insolvency Act 1986 in respect of his company' shall be substituted.

Clause 7.5: Determination by Contractor✽

In clause 7.5.3, delete 'one month', and substitute '[six] months.'

Clause 7.6: Employer becoming bankrupt, etc.✽

In clause 7.6, the words 'or has an application made under the Insolvency Act 1986 in respect of his company to the court for the appointment of an administrator' shall be deleted, and 'or has an administrator appointed by the court under the Insolvency Act 1986 in respect of his company' shall be substituted.

Clause 8.3: Definitions

The definition of 'Joint Names Policy' shall be deleted, and the following substituted:

'Joint Names Policy means a policy of insurance which includes the Contractor and the Employer and such other persons as the Employer may reasonably require (including, without limitation, any persons who have entered or shall enter into an agreement for the provision of finance in connection with the Works, and any persons who have acquired or shall acquire any interest in or over the Works or any part thereof) as the insured.'

Supplementary Condition C: Fluctuations✽

Delete clause C4.8.1.

Supplementary Condition D: Fluctuations✱

Delete clause D13.1.

Additional Conditions

The following clauses shall be added:

'10:   Collateral Warranties✱

10.1   The Contractor shall within 7 (seven) working days of the Employer's request so to do:

10.1.1   execute, in favour of any persons who have entered or shall enter into an agreement for the provision of finance in connection with the Works, a Deed in the form annexed as Annexe 'A', or a similar form reasonably required by the Employer, and deliver the same to the Employer, together with a guarantee in the form annexed as Annexe 'C', or a similar form reasonably required by the Employer, from the ultimate parent company of the Contractor, namely [   ] [Limited] **OR** [PLC], in respect of the Contractor's obligations pursuant to such Deed; and

10.1.2   execute, in favour of any persons who have acquired or shall acquire any interest in or over the Works or any part thereof, a Deed in the form annexed as Annexe 'B', or a similar form reasonably required by the Employer, and deliver the same to the Employer, together with a guarantee in the form annexed as Annexe 'C', or a similar form reasonably required by the Employer, from the ultimate parent company of the Contractor, namely [   ] [Limited] **OR** [PLC], in respect of the Contractor's obligations pursuant to such Deed.

10.2   The Contractor shall (without request) use reasonable endeavours to procure that each sub-contract or supply contract with each person named pursuant to clause 3.3.1 who is to be employed by the Contractor as a sub-contractor, and any other sub-contractor or supplier (if requested by the Employer), shall contain obligations on the relevant sub-contractor or supplier to execute, in favour of the Employer and/or (within 7 (seven) working days of the Employer's request so to do) any persons who have entered or shall enter into an agreement for the provision of finance in connection with the Works and/or in favour of any persons who have acquired or shall acquire any interest in or over the Works or any part thereof, a Deed in the form annexed as Annexe 'D', or a similar form reasonably required by the Employer, and deliver the same to the Employer; together in each case (if requested by the Employer) with a guarantee in the form annexed as Annexe 'E', or a similar form reasonably required by the Employer, from the ultimate parent company of the relevant sub-contractor or supplier in respect of the sub-contractor's or supplier's obligations pursuant to such Deed. The Contractor shall enforce such obligations, or such modified obligations as are referred to below.

10.3   If, despite the Contractor having used such reasonable endeavours, the sub-contractor or supplier will not accept such obligations, or will only accept them in a modified form, the Contractor shall notify the Employer, who may agree in writing that the relevant sub-contract or supply contract need not contain such obligations, or that the relevant obligations may be in a modified form agreeable to the sub-contractor or supplier.

10.4   Failing such agreement by the Employer in the case of a person named pursuant to clause 3.3.1 who is to be employed by the Contractor as a sub-contractor, the Contractor shall not be obliged to enter into a sub-contract or supply contract with that person.

10.5   Failing such agreement by the Employer in the case of any other sub-contractor or supplier, the Contractor shall not enter into a relevant sub-contract or supply contract with that sub-contractor or supplier.

10.6   The above obligations for the provision of Deeds and guarantees in favour of third parties shall continue notwithstanding termination of this Contract, or determination of the Contractor's employment hereunder, in either case for any reason whatsoever, including (without limitation) breach by the Employer. However, any such Deed given after such termination or determination shall be amended by the Employer so as to refer to the fact and date of such termination or determination, to omit any obligation to continue to exercise skill, care or diligence or fulfil obligations pursuant to this Contract after such termination or determination, and to omit any provision enabling a third party to assume the position of employer of the Contractor.

11:   [Performance Bond] OR [Parent Company Guarantee]

Upon execution of this Contract, the Contractor shall deliver to the Employer a performance bond from [   ] [Limited] **OR** [PLC] in an amount of 10% (ten per cent) of the Contract Sum in the form annexed as Annexe 'F' **OR** parent company guarantee from its ultimate parent company, namely [   ] [Limited] **OR** [PLC], for its performance of this Contract in the form annexed as Annexe 'F'.

12: Proscribed Materials

12.1 The Contractor warrants:

    12.1.1 that he has not used or specified and will not use or specify for use;

    12.1.2 that he has exercised and will continue to exercise reasonable skill, care and diligence to see that there are not used;

    12.1.3 that he is not aware and has no reason to suspect or believe that there have been or will be used; and

    12.1.4 that he will promptly notify the Employer in writing if he becomes aware or has reason to suspect or believe that there have been or will be used;

in or in connection with the Works, any of the materials or substances identified in clause 12.2.

12.2 The said materials or substances are:

    12.2.1 high alumina cement in structural elements;

    12.2.2 wood wool slabs in permanent formwork to concrete;

    12.2.3 calcium chloride in admixtures for use in reinforced concrete;

    12.2.4 asbestos products;

    12.2.5 naturally occurring aggregates for use in reinforced concrete which do not comply with British Standard 882: 1983 and/or naturally occurring aggregates for use in concrete which do not comply with British Standard 8110: 1985.'

APPENDIX*

ANNEXE 'F'∗

DATED   **199**

[   ]

('the Contractor')

-and-

[   ]

('the Surety')

-and-

[   ]

('the Employer')

PERFORMANCE BOND

in respect of the construction of
[   ]

THIS BOND is made the   day of   One thousand nine hundred and ninety-

BETWEEN:

(1)

[of] **OR** [whose registered office is at]

('the Contractor');

(2)

[of] **OR** [whose registered office is at]

('the Surety'); and

(3)

[of] **OR** [whose registered office is at]

('the Employer', which term shall include its successors and assigns).

WHEREAS by an Agreement ('the Contract') dated   199   and made between the Employer of the one part and the Contractor of the other part, the Contractor undertook the construction of certain Works in accordance with the terms and conditions of the Contract.

NOW THIS DEED WITNESSETH as follows:

1   Bond

By this Bond the Contractor and the Surety, their successors and assigns, are jointly and severally held and bound to the Employer for payment to the Employer of the sum of   pounds (£   ).

2   Conditions

The conditions of this Bond are that if:

2.1   the Contractor duly discharges all the Contractor's obligations under or pursuant to the Contract; or

2.2   in the event of the Contractor's default in the discharge of any such obligations, the Surety shall pay to the Employer the loss and damage thereby caused to the Employer, up to the amount of this Bond; or

2.3   pursuant to clause 4.6 of the Contract, any balance due from the Contractor to the Employer pursuant to a Final Certificate has been paid; and any amount due from the Contractor to the Employer pursuant to any award or judgment in, or settlement of, any arbitration or other proceedings commenced in respect of the Contract before or within 28 (twenty-eight) days after the said Final Certificate was issued, has been paid;

this Bond shall thereby be discharged, but otherwise shall remain in force.

3   Alterations

No alterations in the Contract, or in the Works, and no extension of time, forbearance or forgiveness, nor any act, matter or thing whatsoever except fulfilment of one of the above conditions or an express release by Deed by the Employer, shall in any way release the Surety from any liability under this Bond.

[4   Reduction on Practical Completion]

[Provided that upon certification of Practical Completion of the Works under the Contract, the amount of this Bond shall reduce by one-half] **OR** [Provided that upon certification of Practical Completion of each Section of the Works under the Contract, the amount of this Bond shall reduce by the following respective amounts:

£

Section 1

Section 2   ]

IN WITNESS whereof the Contractor and the Surety have executed this Deed on the date first stated above.

COMMENTARY

Article 3: Architect or Contract Administrator
Article 4: Quantity Surveyor

See paragraph 5, Chapter 2.

Clause 1.3: Priority of Contract Documents

See paragraph 7, Chapter 2.

Clause 2.4: Events referred to in 2.3

See paragraphs 11-14, Chapter 2.

Clause 2.7: Liquidated damages for non-completion

See paragraph 10, Chapter 2.

Clause 3.1: Assignment

See paragraphs 8 and 9, Chapter 2.

Clause 3.3: Named persons as sub-contractors

See paragraph 26, Chapter 1, and the commentary upon Form 12 relating to Nominated Sub-Contractors.

Clause 4.2: Interim Payment

See paragraph 23, Chapter 2.

Clause 4.4: Interest in percentage withheld

See paragraph 22, Chapter 2.

Clause 7.1: Determination by Employer

See paragraph 16, Chapter 2.

Clause 7.2: Contractor becoming bankrupt, etc.

See paragraph 15, Chapter 2.

Clause 7.5: Determination by Contractor

See commentary relating to the Appendix to Form 12.

Clause 7.6: Employer becoming bankrupt, etc.

See paragraph 15, Chapter 2.

Supplementary Conditions C and D: Fluctuations

See paragraph 25, Chapter 2.

Clause 10: Collateral Warranties

For Annexes 'A', 'B', 'C', 'D' and 'E' see Forms 19, 20, 28, 21 and 29.

Appendix

The published Appendix should be completed and attached.

Annexe 'F': Performance Bond or Parent Company Guarantee

Only a performance bond is given as Annexe 'F' to this Form. Cf. Form 30. If a parent company guarantee is required, Form 31 may be used.

FORM 18: MAIN CONTRACT INCORPORATING JCT AGREEMENT FOR MINOR WORKS 1980, INCLUDING PERFORMANCE BOND

DATED _____ **199**

[   ]

('the Employer')

-and-

[   ]

('the Contractor')

MAIN CONTRACT

for the construction of

[   ]

✱ An asterisk indicates that there is a relevant note in the commentary upon this Form.

THIS AGREEMENT made the　　　　　　　day of　　　　　　　One thousand nine hundred and ninety-

BETWEEN:

(1)

[of] **OR** [whose registered office is at]

('the Employer', which term shall include its successors and assigns); and

(2)

[of] **OR** [whose registered office is at]

('the Contractor');

INCORPORATES the Joint Contracts Tribunal Agreement for Minor Building Works, 1980 Edition, as printed in　　　　　　　199　, incorporating Amendments MW1 (April 1985), MW2 (November 1986), MW3 (August 1987), MW4 (October 1987), MW5 (July 1988), MW6 (April 1989) and MW7 (July 1991), and as amended by the Schedule.

IN WITNESS whereof the parties have executed this Deed in duplicate on the date first stated above.

SCHEDULE

Recitals

First recital

The work is the construction of

at

and is to be carried out under the direction of

[of] **OR** [whose registered office is at]

(hereinafter called ['the Architect'] **OR** ['the Contract Administrator']).

The Contract Documents are:

[– Contract Drawings numbered:

;]

[– a Contract Specification; and]

[– Schedules numbered:

.]

Fourth recital

The name and address of the Quantity Surveyor are

[of] **OR** [whose registered office is at]

Fifth recital✳

Articles of Agreement

Article 2

The sum to be inserted is   pounds (£   ).

Article 3

The name and address of the [Architect] **OR** [Contract Administrator] are

[of] **OR** [whose registered office is at]

Article 4✱

The appointing institution shall be the [Royal Institute of British Architects] **OR** [Royal Institution of Chartered Surveyors] **OR** [Chartered Institute of Arbitrators].

Amendments to Conditions

Clause 2.0: Commencement and completion

Commencement and completion

Delete clause 2.1, and substitute:

'2.1   The Works may be commenced on   199   and shall be completed
by   199   .'

Extension of contract period✱

In clause 2.2, add after 'including':

> 'any breach of this Contract by the Employer or any act or omission on the part of the Employer, the [Architect] **OR** [Contract Administrator] or the Employer's Quantity Surveyor and ...'.

Damages for non-completion

In clause 2.3, insert:

'   pounds (£   ) per day.'

Defects liability

In clause 2.5, delete 'three months', and substitute '[   ] months'.

Clause 3.0: Control of the Works

Assignment✱

Delete clause 3.1, and substitute:

'3.1   The Contractor shall not assign this Contract. The Employer may assign this Contract.'

Clause 4.0: Payment

Correction of inconsistencies✱

In clause 4.1, delete the second sentence.

Final certificate

In clause 4.4, delete 'three months', and substitute '[   ] months'.

Contribution, levy and tax charges

[Clause 4.5 shall be deleted.]

OR

[In clause 4.5, the percentage addition under Part A, clause A5 shall be   per cent
(%).]

Clause 5.0: Statutory obligations

Value Added tax

In clause 5.2, clause B1.1 of the Supplementary Memorandum, Part B, [applies] **OR** [does not apply].

Clause 6.0: Injury, damage and insurance

Injury or damage to property

In clause 6.2, the sum to be inserted shall be   pounds (£   ).

[Insurance of the Works – Fire, etc. – New Works

Clause 6.3A shall apply, and clause 6.3B shall be deleted. The percentage to be inserted in clause 6.3A shall be [fifteen per cent (15%)].]

OR

[Insurance of the Works – Fire, etc. – Existing Structures

Clause 6.3B shall apply, and clause 6.3A shall be deleted.]

Clause 7.0: Determination

Determination by Employer✱

In clause 7.1, delete 'but not unreasonably or vexatiously'.

In clause 7.2.2, the words 'or has an application made under the Insolvency Act 1986 in respect of his company to the court for the appointment of an administrator' shall be deleted, and 'or has an administrator appointed by the court under the Insolvency Act 1986 in respect of his company' shall be substituted.

Determination by Contractor✱

In clause 7.2.3, delete 'one month' and substitute '[six] months'.

In clause 7.2.4, the words 'or has an application made under the Insolvency Act 1986 in respect of his company to the court for the appointment of an administrator' shall be deleted, and 'or has an administrator appointed by the court under the Insolvency Act 1986 in respect of his company' shall be substituted.

[Clause 8.0: Supplementary Memorandum

Delete 'the 5th recital and'.]

Supplementary Memorandum: Part A: Fluctuations***

Delete clause A4.4.4.2.1.

Additional Conditions

The following clauses shall be added:

'10.0:   Collateral Warranties***

10.1   The Contractor shall within 7 (seven) working days of the Employer's request so to do:

10.1.1   execute, in favour of any persons who have entered or shall enter into an agreement for the provision of finance in connection with the Works, a Deed in the form annexed as Annexe 'A', or a similar form reasonably required by the Employer, and deliver the same to the Employer, together with a guarantee in the form annexed as Annexe 'C', or a similar form reasonably required by the Employer, from the ultimate parent company of the Contractor, namely [   ] [Limited] **OR** [PLC], in respect of the Contractor's obligations pursuant to such Deed; and

10.1.2   execute, in favour of any persons who have acquired or shall acquire any interest in or over the Works or any part thereof, a Deed in the form annexed as Annexe 'B', or a similar form reasonably required by the Employer, and deliver the same to the Employer, together with a guarantee in the form annexed as Annexe 'C', or a similar form reasonably required by the Employer, from the ultimate parent company of the Contractor, namely [   ] [Limited] **OR** [PLC], in respect of the Contractor's obligations pursuant to such Deed.

10.2   The above obligations for the provision of Deeds and guarantees in favour of third parties shall continue notwithstanding termination of this Contract, or determination of the Contractor's employment hereunder, in either case for any reason whatsoever, including (without limitation) breach by the Employer. However, any such Deed given after such termination or determination shall be amended by the Employer so as to refer to the fact and date of such termination or determination, to omit any obligation to continue to exercise skill, care or diligence or fulfil obligations pursuant to this Contract after such termination or determination, and to omit any provision enabling a third party to assume the position of employer of the Contractor.

11.0:   [Performance Bond] OR [Parent Company Guarantee]

Upon execution of this Contract, the Contractor shall deliver to the Employer a performance bond from [   ] [Limited] **OR** [PLC] in an amount of 10% (ten per cent) of the Contract Sum in the form annexed as Annexe 'D' **OR** parent company guarantee from its ultimate parent company, namely [   ] [Limited] **OR** [PLC], for its performance of this Contract in the form annexed as Annexe 'D'.

12.0:   Proscribed Materials

12.1   The Contractor warrants:

    12.1.1   that he has not used or specified and will not use or specify for use;

    12.1.2   that he has exercised and will continue to exercise reasonable skill, care and diligence to see that there are not used;

    12.1.3   that he is not aware and has no reason to suspect or believe that there have been or will be used; and

    12.1.4   that he will promptly notify the Employer in writing if he becomes aware or has reason to suspect or believe that there have been or will be used;

in or in connection with the Works, any of the materials or substances identified in clause 12.2.

12.2   The said materials or substances are:

    12.2.1   high alumina cement in structural elements;

    12.2.2   wood wool slabs in permanent formwork to concrete;

    12.2.3   calcium chloride in admixtures for use in reinforced concrete;

    12.2.4   asbestos products;

12.2.5   naturally occurring aggregates for use in reinforced concrete which do not comply with British Standard 882: 1983 and/or naturally occurring aggregates for use in concrete which do not comply with British Standard 8110: 1985.'

ANNEXE 'D'*

DATED   **199**

[   ]

('the Contractor')

-and-

[   ]

('the Surety')

-and-

[   ]

('the Employer')

PERFORMANCE BOND

in respect of the construction of

[   ]

THIS BOND is made the   day of   One thousand nine hundred and ninety-

BETWEEN:

(1)

        [of] **OR** [whose registered office is at]

        ('the Contractor');

(2)

        [of] **OR** [whose registered office is at]

        ('the Surety'); and

(3)

        [of] **OR** [whose registered office is at]

        ('the Employer', which term shall include its successors and assigns).

WHEREAS by an Agreement ('the Contract') dated   199   and made between the Employer of the one part and the Contractor of the other part, the Contractor undertook the construction of certain Works in accordance with the terms and conditions of the Contract.

NOW THIS DEED WITNESSETH as follows:

1   Bond

By this Bond the Contractor and the Surety, their successors and assigns, are jointly and severally held and bound to the Employer for payment to the Employer of the sum of   pounds (£   ).

2   Conditions

The conditions of this Bond are that if:

2.1   the Contractor duly discharges all the Contractor's obligations under or pursuant to the Contract; or

2.2   in the event of the Contractor's default in the discharge of any such obligations, the Surety shall pay to the Employer the loss and damage thereby caused to the Employer, up to the amount of this Bond; or

2.3   pursuant to clause 4.4 of the Contract, any balance due from the Contractor to the Employer pursuant to a Final Certificate has been paid; and any amount due from the Contractor to the Employer pursuant to any award or judgment in, or settlement of, any arbitration or other proceedings commenced in respect of the Contract before or within 28 (twenty-eight) days after the said Final Certificate was issued, has been paid;

this Bond shall thereby be discharged, but otherwise shall remain in force.

3   Alterations

No alterations in the Contract, or in the Works, and no extension of time, forbearance or forgiveness, nor any act, matter or thing whatsoever except fulfilment of one of the above conditions or an express release by Deed by the Employer, shall in any way release the Surety from any liability under this Bond.

[4   Reduction on Practical Completion

Provided that upon certification of Practical Completion of the Works under the Contract, the amount of this Bond shall reduce by one-half.]

IN WITNESS whereof the Contractor and the Surety have executed this Deed on the date first stated above.

COMMENTARY

Fifth recital

The Building Employers' Confederation ('BEC') Guarantee Scheme operated by the BEC Building Trust Limited, referred to in Part E, Section 1 of the Supplementary Memorandum, is no longer operative. If some other equivalent scheme is to be used, the recital should be completed accordingly.

Article 4

It is recommended that the Chartered Institute of Arbitrators shall be selected, in view of that Institute's training of its arbitrators.

Clause 2.2: Extension of contract period

See paragraph 13, Chapter 2.

Clause 3.1: Assignment

See paragraphs 8 and 9, Chapter 2.

Clause 4.1: Correction of inconsistencies

See paragraph 7, Chapter 2.

Clause 7.0: Determination by Employer

See paragraphs 15 and 16, Chapter 2.

Clause 7.0: Determination by Contractor

See commentary relating to the Appendix to Form 12.

Supplementary Memorandum: Part A: Fluctuations

See paragraph 25, Chapter 2.

Clause 10.0: Collateral Warranties

Collateral warranties may well not be necessary at all under Minor Works Contracts, and collateral warranties from sub-contractors and suppliers are not provided for. For Annexes 'A', 'B' and 'C' see Forms 19, 20 and 28.

Annexe 'D': Performance Bond or Parent Company Guarantee

Only a performance bond is given as Annexe 'C' to this Form. Cf. Form 30. Sectional completion is not provided for in the Minor Works documentation, and therefore Annexe 'D' does not provide for reduction of the bond on sectional completion. If a parent company guarantee is required, Form 31 may be used.

FORM 19: COLLATERAL WARRANTY BY CONTRACTOR TO FUNDING INSTITUTION, FOR USE AS AN ANNEXE TO FORMS 12–18

ANNEXE 'A'

DATED _____ **199**

[   ]

('the Contractor')

-and-

[   ]

('the Employer')

-and-

[   ]

('the Fund')

COLLATERAL WARRANTY BY CONTRACTOR TO FUNDING INSTITUTION

in respect of the construction of

[   ]

✶ An asterisk indicates that there is a relevant note in the commentary upon this Form.

THIS AGREEMENT is made the   day of   One thousand nine hundred and ninety-

BETWEEN:

(1)

[of] **OR** [whose registered office is at]

('the Contractor');

(2)

[of] **OR** [whose registered office is at]

('the Employer'); and

(3)

[whose registered office is at]

('the Fund', which term shall include its successors and assigns).

WHEREAS:

(A)   The Fund has entered into an agreement ('the Finance Agreement') with the Employer for the provision of certain finance in connection with the carrying out of a project of development briefly described as

          at

                                                                              ('the Development').

          [The Fund entered into the Finance Agreement, and enters into this Agreement, on its own behalf and as agent for a syndicate of banks. Each of the banks which are members of the syndicate from time to time, including banks joining the syndicate after the date of this Agreement, shall be entitled to the benefit of this Agreement in addition to the Fund.]

(B)   The Employer has entered into a building contract dated   199   ('the Building Contract') with the Contractor for the construction of the Development.

NOW in consideration of £1 (one pound) paid by the Fund to the Contractor (receipt of which the Contractor hereby acknowledges) **THIS DEED WITNESSETH** as follows:

1.✱   The Contractor covenants with the Fund that it has duly performed and observed, and will continue duly to perform and observe, all the terms of the Building Contract on the Contractor's part to be performed and observed and, without prejudice to the generality of the foregoing, the Contractor warrants that it has exercised and will continue to exercise reasonable skill, care and diligence in the performance of its duties to the Employer under the Building Contract.

2.1   Without prejudice to the generality of clause 1, the Contractor further warrants:

   2.1.1   that it has not used or specified and will not use or specify for use;

   2.1.2   that it has exercised and will continue to exercise reasonable skill, care and diligence to see that there are not used;

   2.1.3   that it is not aware and has no reason to suspect or believe that there have been or will be used; and

   2.1.4   that it will promptly notify the Fund in writing if it becomes aware or has reason to suspect or believe that there have been or will be used;

   in or in connection with the Development, any of the materials or substances identified in clause 2.2.

2.2   The said materials or substances are:

   2.2.1   high alumina cement in structural elements;

   2.2.2   wood wool slabs in permanent formwork to concrete;

   2.2.3   calcium chloride in admixtures for use in reinforced concrete;

   2.2.4   asbestos products;

   2.2.5   naturally occurring aggregates for use in reinforced concrete which do not comply with British Standard 882: 1983 and/or naturally occurring aggregates for use in concrete which do not comply with British Standard 8110: 1985.

3.   The Fund has no authority to issue any direction or instruction to the Contractor in relation to performance of the Contractor's duties under the Building Contract unless

and until the Fund has given notice under clauses 5 or 6.

4.   The Contractor acknowledges that the Employer has paid all sums due and owing to the Contractor under the Building Contract up to the date of this Agreement. The Fund has no liability to the Contractor in respect of sums due under the Building Contract unless and until the Fund has given notice under clauses 5 or 6.

5.   The Contractor agrees that, in the event of the termination of the Finance Agreement by the Fund, or the occurrence of any event of default under the Finance Agreement, the Contractor will, if so required by notice in writing given by the Fund and subject to clause 7, accept the instructions of the Fund or its appointee to the exclusion of the Employer in respect of the Development upon the terms and conditions of the Building Contract. The Employer acknowledges that the Contractor shall be entitled to rely on a notice given to the Contractor by the Fund under this clause 5 as conclusive evidence for the purposes of this Agreement of the termination of the Finance Agreement by the Fund, or the occurrence of any such event of default.

6.   The Contractor further agrees that it will not, without first giving the Fund not less than 21 (twenty-one) days' notice in writing, exercise any right it may have to terminate the Building Contract or determine its employment thereunder, or to treat the same as having been repudiated by the Employer, or discontinue the performance of any duties to be performed by the Contractor pursuant thereto for any reason whatsoever (including, without limitation, any act or omission by or on behalf of the Employer). The Contractor's right to terminate the Building Contract or determine its employment thereunder, or treat the same as having been repudiated, or discontinue performance shall cease if, within such period of notice and subject to clause 7, the Fund shall give notice in writing to the Contractor requiring the Contractor to accept the instructions of the Fund or its appointee to the exclusion of the Employer in respect of the Development upon the terms and conditions of the Building Contract.

7.   It shall be a condition of any notice given by the Fund under clauses 5 or 6 that the Fund or its appointee accepts liability for payment of all sums payable to the Contractor under the Building Contract and for performance of the Employer's obligations under the Building Contract, including payment of any sums outstanding at the date of such notice, but excluding any such sums in respect of which funds were advanced to the Employer pursuant to the Finance Agreement before the date of such notice. Upon the issue of any notice by the Fund under clauses 5 or 6, the Building Contract shall continue in full force and effect as if no right of termination or determination on the part of the Contractor had arisen and the Contractor shall be liable to the Fund or its appointee under the Building Contract in lieu of its liability to the Employer. If any notice given by the Fund under clauses 5 or 6 requires the Contractor to accept the instructions of the Fund's appointee, the Fund shall be liable to the Contractor as

guarantor for the payment of all sums from time to time due to the Contractor from the Fund's appointee.

8.✱   The copyright in all drawings, reports, specifications, bills of quantities, calculations and other similar documents provided by the Contractor in connection with the Development shall remain vested in the Contractor, but the Fund and its appointee shall have a licence to copy and use such drawings and other documents, and to reproduce the designs contained in them, for any purpose related to the Development including, but without limitation, the construction, completion, maintenance, letting, promotion, advertisement, reinstatement, repair and/or extension of the Development. The Contractor shall, if the Fund so requests and undertakes in writing to pay the Contractor's reasonable copying charges, promptly supply the Fund with conveniently reproducible copies of all such drawings and other documents.

9.1✱   The Contractor shall maintain professional indemnity insurance covering (inter alia) all liability hereunder upon customary and usual terms and conditions prevailing for the time being in the insurance market, and with reputable insurers lawfully carrying on such insurance business in the United Kingdom, in an amount of not less than   pounds (£   ) for any one occurrence or series of occurrences arising out of any one event for a period beginning now and ending 15 (fifteen) years after the date of practical completion of the Development for the purposes of the Building Contract, provided always that such insurance is available at commercially reasonable rates. The said terms and conditions shall not include any term or condition to the effect that the Contractor must discharge any liability before being entitled to recover from the insurers, or any other term or condition which might adversely affect the rights of any person to recover from the insurers pursuant to the Third Parties (Rights Against Insurers) Act 1930, or any amendment or re-enactment thereof. The Contractor shall not, without the prior approval in writing of the Fund, settle or compromise with the insurers any claim which the Contractor may have against the insurers and which relates to a claim by the Fund against the Contractor, or by any act or omission lose or prejudice the Contractor's right to make or proceed with such a claim against the insurers.

9.2   Any increased or additional premium required by insurers by reason of the Contractor's own claims record or other acts, omissions, matters or things particular to the Contractor shall be deemed to be within commercially reasonable rates.

9.3   The Contractor shall immediately inform the Fund if such insurance ceases to be available at commercially reasonable rates in order that the Contractor and the Fund can discuss means of best protecting the respective positions of the Fund and the Contractor in respect of the Development in the absence of such insurance.

9.4   The Contractor shall fully co-operate with any measures reasonably required by the Fund, including (without limitation) completing any proposals for insurance and associated documents, maintaining such insurance at rates above commercially reasonable rates if the Fund undertakes in writing to reimburse the Contractor in respect of the net cost of such insurance to the Contractor above commercially reasonable rates or, if the Fund effects such insurance at rates at or above commercially reasonable rates, reimbursing the Fund in respect of what the net cost of such insurance to the Fund would have been at commercially reasonable rates.

9.5   As and when it is reasonably requested to do so by the Fund or its appointee under clauses 5 or 6, the Contractor shall produce for inspection documentary evidence (including, if required by the Fund or such appointee, the original of the relevant insurance documents) that its professional indemnity insurance is being maintained.

10.   The Employer has agreed to be a party to this Agreement for the purpose of clause 12 and for acknowledging that the Contractor shall not be in breach of the Building Contract by complying with the obligations imposed on it by this Agreement.

11.   This Agreement may be assigned by the Fund and its successors and assigns without the consent of the Employer or the Contractor being required.

12.   The Employer and the Contractor undertake with the Fund not to vary, or depart from, the terms and conditions of the Building Contract without the prior written consent of the Fund, and agree that no such variation or departure made without such consent shall be binding on the Fund, or affect or prejudice the Fund's rights hereunder, or under the Building Contract, or in any other way.

13.   Any notice to be given by the Contractor hereunder shall be deemed to be duly given if it is delivered by hand at or sent by registered post or recorded delivery to the Fund at its registered office, and any notice to be given by the Fund hereunder shall be deemed to be duly given if it is addressed to the Contractor and delivered by hand at or sent by registered post or recorded delivery to the above-mentioned address of the Contractor or to the principal business address of the Contractor for the time being and, in the case of any such notices, the same shall if sent by registered post or recorded delivery be deemed to have been received forty-eight hours after being posted.

IN WITNESS whereof the Contractor and the Employer have executed this Deed on the date first stated above.

COMMENTARY

See commentary upon Form 6.

Clause 1

This clause goes beyond a duty of care, as it contains a covenant actually to perform the Building Contract.

Clauses 8 and 9

These clauses may be irrelevant unless the Contractor has assumed design responsibility under the Building Contract.

FORM 20: COLLATERAL WARRANTY BY CONTRACTOR TO ACQUIRER, FOR USE AS AN ANNEXE TO FORMS 12–18

ANNEXE 'B'

DATED _____ **199**

[   ]

('the Contractor')

-and-

[   ]

('the Employer')

-and-

[   ]

('the Acquirer')

COLLATERAL WARRANTY BY CONTRACTOR TO ACQUIRER

in respect of the construction of

[   ]

✱ An asterisk indicates that there is a relevant note in the commentary upon this Form.

THIS AGREEMENT is made the   day of   One thousand nine hundred and ninety-

BETWEEN:

(1)

       [of] **OR** [whose registered office is at]

       ('the Contractor');

(2)

       [of] **OR** [whose registered office is at]

       ('the Employer'); and

(3)

       [of] **OR** [whose registered office is at]

       ('the Acquirer', which term shall include its successors and assigns).

WHEREAS:

(A)   The Acquirer intends to acquire, or has acquired, an interest in a project of development briefly described as

       at

                                     ('the Development').

(B)   The Employer has entered into a building contract dated   199
       ('the Building Contract') with the Contractor for the construction of the Development.

NOW in consideration of £1 (one pound) paid by the Acquirer to the Contractor (receipt of which the Contractor hereby acknowledges) **THIS DEED WITNESSETH** as follows:

1.✱   The Contractor covenants with the Acquirer that it has duly performed and observed, and will continue duly to perform and observe, all the terms of the Building Contract on the Contractor's part to be performed and observed and, without prejudice to the generality of the foregoing, the Contractor warrants that it has exercised and will continue to exercise reasonable skill, care and diligence in the performance of its duties to the Employer under the Building Contract.

2.1   Without prejudice to the generality of clause 1, the Contractor further warrants:

     2.1.1   that it has not used or specified and will not use or specify for use;

     2.1.2   that it has exercised and will continue to exercise reasonable skill, care and diligence to see that there are not used;

     2.1.3   that it is not aware and has no reason to suspect or believe that there have been or will be used; and

     2.1.4   that it will promptly notify the Acquirer in writing if it becomes aware or has reason to suspect or believe that there have been or will be used;

in or in connection with the Development, any of the materials or substances identified in clause 2.2.

2.2   The said materials or substances are:

     2.2.1   high alumina cement in structural elements;

     2.2.2   wood wool slabs in permanent formwork to concrete;

     2.2.3   calcium chloride in admixtures for use in reinforced concrete;

     2.2.4   asbestos products;

     2.2.5   naturally occurring aggregates for use in reinforced concrete which do not comply with British Standard 882: 1983 and/or naturally occurring aggregates for use in concrete which do not comply with British Standard 8110: 1985.

3.   The Acquirer has no authority to issue any direction or instruction to the Contractor in relation to performance of the Contractor's duties under the Building Contract.

4.   The Contractor acknowledges that the Employer has paid all sums due and owing to the Contractor under the Building Contract up to the date of this Agreement. The Acquirer has no liability to the Contractor in respect of sums due under the Building Contract.

5.✱   The copyright in all drawings, reports, specifications, bills of quantities, calculations and other similar documents provided by the Contractor in connection with the Development shall remain vested in the Contractor, but the Acquirer and its appointee shall have a licence to copy and use such drawings and other documents, and to reproduce the designs contained in them, for any purpose related to the Development including, but without limitation, the construction, completion, maintenance, letting, promotion, advertisement, reinstatement, repair and/or extension of the Development. The Contractor shall, if the Acquirer so requests and undertakes in writing to pay the Contractor's reasonable copying charges, promptly supply the Acquirer with conveniently reproducible copies of all such drawings and other documents.

6.1✱ The Contractor shall maintain professional indemnity insurance covering (inter alia) all liability hereunder upon customary and usual terms and conditions prevailing for the time being in the insurance market, and with reputable insurers lawfully carrying on such insurance business in the United Kingdom, in an amount of not less than   pounds (£   ) for any one occurrence or series of occurrences arising out of any one event for a period beginning now and ending 15 (fifteen) years after the date of practical completion of the Development for the purposes of the Building Contract, provided always that such insurance is available at commercially reasonable rates. The said terms and conditions shall not include any term or condition to the effect that the Contractor must discharge any liability before being entitled to recover from the insurers, or any other term or condition which might adversely affect the rights of any person to recover from the insurers pursuant to the Third Parties (Rights Against Insurers) Act 1930, or any amendment or re-enactment thereof. The Contractor shall not, without the prior approval in writing of the Acquirer, settle or compromise with the insurers any claim which the Contractor may have against the insurers and which relates to a claim by the Acquirer against the Contractor, or by any act or omission lose or prejudice the Contractor's right to make or proceed with such a claim against the insurers.

6.2 Any increased or additional premium required by insurers by reason of the Contractor's own claims record or other acts, omissions, matters or things particular to the Contractor shall be deemed to be within commercially reasonable rates.

6.3 The Contractor shall immediately inform the Acquirer if such insurance ceases to be available at commercially reasonable rates in order that the Contractor and the Acquirer can discuss means of best protecting the respective positions of the Acquirer and the Contractor in respect of the Development in the absence of such insurance.

6.4 The Contractor shall fully co-operate with any measures reasonably required by the Acquirer, including (without limitation) completing any proposals for insurance and associated documents, maintaining such insurance at rates above commercially reasonable rates if the Acquirer undertakes in writing to reimburse the Contractor in respect of the net cost of such insurance to the Contractor above commercially reasonable rates or, if the Acquirer effects such insurance at rates at or above commercially reasonable rates, reimbursing the Acquirer in respect of what the net cost of such insurance to the Acquirer would have been at commercially reasonable rates.

6.5   As and when it is reasonably requested to do so by the Acquirer the Contractor shall produce for inspection documentary evidence (including, if required by the Acquirer, the original of the relevant insurance documents) that its professional indemnity insurance is being maintained.

7.   The Employer has agreed to be a party to this Agreement for the purpose of clause 9 and for acknowledging that the Contractor shall not be in breach of the Building Contract by complying with the obligations imposed on it by this Agreement.

8.   This Agreement may be assigned by the Acquirer and its successors and assigns without the consent of the Employer or the Contractor being required.

9.   The Employer and the Contractor undertake with the Acquirer not to vary, or depart from, the terms and conditions of the Building Contract without the prior written consent of the Acquirer, and agree that no such variation or departure made without such consent shall be binding on the Acquirer, or affect or prejudice the Acquirer's rights hereunder, or in any other way.

10.   Any notice to be given by the Contractor hereunder shall be deemed to be duly given if it is delivered by hand at or sent by registered post or recorded delivery to the above-mentioned address of the Acquirer or to the principal business address of the Acquirer for the time being, and any notice to be given by the Acquirer hereunder shall be deemed to be duly given if it is addressed to the Contractor and delivered by hand at or sent by registered post or recorded delivery to the above-mentioned address of the Contractor or to the principal business address of the Contractor for the time being and, in the case of any such notices, the same shall if sent by registered post or recorded delivery be deemed to have been received forty-eight hours after being posted.

IN WITNESS whereof the Contractor and the Employer have executed this Deed on the date first stated above.

COMMENTARY

See commentaries upon Forms 6, 7 and 19.

Clause 1

This clause goes beyond a duty of care, as it contains a covenant actually to perform the Building Contract.

Clauses 5 and 6

These clauses may be irrelevant unless the Contractor has assumed design responsibility under the Building Contract.

FORM 21: COLLATERAL WARRANTY BY SUB-CONTRACTOR, FOR USE AS AN ANNEXE TO FORMS 12–18

ANNEXE 'D'

DATED _____ **199**

[   ]

('the Sub-Contractor')

-and-

[   ]

(['the Employer'] **OR** ['the Fund'] **OR** ['the Acquirer'])

COLLATERAL WARRANTY BY SUB-CONTRACTOR

in respect of the construction of
[   ] .

✱ An asterisk indicates that there is a relevant note in the commentary upon this Form.

THIS AGREEMENT is made the   day of   One thousand nine hundred and ninety-

BETWEEN:

(1)

[of] **OR** [whose registered office is at]

('the Sub-Contractor'); and

(2)

[of] **OR** [whose registered office is at]

(['the Employer'] **OR** ['the Fund'] **OR** ['the Acquirer'], which term shall include its successors and assigns).

WHEREAS:

(A)   [The Employer] **OR** [('the Employer')] has entered into a building contract dated   199   ('the Building Contract') with [   ] ('the Contractor') for the construction of a project of development briefly described as

at

                                                                                        ('the Development').

(B)   The Contractor has entered into a sub-contract dated   199   ('the Sub-Contract') with the Sub-Contractor for the execution of certain sub-contract works briefly described as

and forming part of the Development ('the Sub-Contract Works').

[(C)   The Fund has entered into an agreement with the Employer for the provision of certain finance in connection with the carrying out of the Development.] [The Fund entered into such agreement, and enters into this Agreement, on its own behalf and as agent for a syndicate of banks. Each of the banks which are members of the syndicate from time to time, including banks joining the syndicate after the date of this Agreement, shall be entitled to the benefit of this Agreement in addition to the Fund.]

OR

[(C)   The Acquirer intends to acquire, or has acquired, an interest in the Development.]

NOW in consideration of £1 (one pound) paid by the [Employer] **OR** [Fund] **OR** [Acquirer] to the Sub-Contractor (receipt of which the Sub-Contractor hereby acknowledges) **THIS DEED WITNESSETH** as follows:

1.✱   The Sub-Contractor covenants with the [Employer] **OR** [Fund] **OR** [Acquirer] that it has duly performed and observed, and will continue duly to perform and observe, all the terms of the Sub-Contract on the Sub-Contractor's part to be performed and observed and, without prejudice to the generality of the foregoing, the Sub-Contractor warrants that it has exercised and will continue to exercise reasonable skill, care and diligence in the performance of its duties to the Contractor under the Sub-Contract.

2.1   Without prejudice to the generality of clause 1, the Sub-Contractor further warrants:

   2.1.1   that it has not used or specified and will not use or specify for use;

   2.1.2   that it has exercised and will continue to exercise reasonable skill, care and diligence to see that there are not used;

   2.1.3   that it is not aware and has no reason to suspect or believe that there have been or will be used; and

   2.1.4   that it will promptly notify the [Employer] **OR** [Fund] **OR** [Acquirer] in writing if it becomes aware or has reason to suspect or believe that there have been or will be used;

   in or in connection with the Development, any of the materials or substances identified in clause 2.2.

2.2   The said materials or substances are:

   2.2.1   high alumina cement in structural elements;

   2.2.2   wood wool slabs in permanent formwork to concrete;

   2.2.3   calcium chloride in admixtures for use in reinforced concrete;

   2.2.4   asbestos products;

2.2.5   naturally occurring aggregates for use in reinforced concrete which do not comply with British Standard 882: 1983 and/or naturally occurring aggregates for use in concrete which do not comply with British Standard 8110: 1985.

3.   The [Employer] **OR** [Fund] **OR** [Acquirer] has no authority to issue any direction or instruction to the Sub-Contractor in relation to performance of the Sub-Contractor's duties under the Sub-Contract.

4.   The Sub-Contractor acknowledges that the Contractor has paid all sums due and owing to the Sub-Contractor under the Sub-Contract up to the date of this Agreement. The [Employer] **OR** [Fund] **OR** [Acquirer] has no liability to the Sub-Contractor in respect of sums due under the Sub-Contract.

5.✳   The copyright in all drawings, reports, specifications, bills of quantities, calculations and other similar documents provided by the Sub-Contractor in connection with the Development shall remain vested in the Sub-Contractor (or as may be otherwise provided by the Sub-Contract), but the [Employer] **OR** [Fund] **OR** [Acquirer] and its appointee shall have a licence to copy and use such drawings and other documents, and to reproduce the designs contained in them, for any purpose related to the Development including, but without limitation, the construction, completion, maintenance, letting, promotion, advertisement, reinstatement, repair and/or extension of the Development. The Sub-Contractor shall, if the [Employer] **OR** [Fund] **OR** [Acquirer] so requests and undertakes in writing to pay the Sub-Contractor's reasonable copying charges, promptly supply the [Employer] **OR** [Fund] **OR** [Acquirer] with conveniently reproducible copies of all such drawings and other documents.

6.1✳   The Sub-Contractor shall maintain professional indemnity insurance covering (inter alia) all liability hereunder upon customary and usual terms and conditions prevailing for the time being in the insurance market, and with reputable insurers lawfully carrying on such insurance business in the United Kingdom, in an amount of not less than   pounds (£   ) for any one occurrence or series of occurrences arising out of any one event for a period beginning now and ending 15 (fifteen) years after the date of practical completion of the Development for the purposes of the Building Contract, provided always that such insurance is available at commercially reasonable rates. The said terms and conditions shall not include any term or condition to the effect that the Sub-Contractor must discharge any liability before being entitled to recover from the insurers, or any other term or condition which might adversely affect the rights of any person to recover from the insurers pursuant to the Third Parties (Rights Against Insurers) Act 1930, or any amendment or re-enactment thereof. The Sub-Contractor shall not, without the prior approval in writing of the [Employer] **OR** [Fund] **OR** [Acquirer], settle or compromise with the insurers any claim which the Sub-Contractor may have against the insurers and which relates to a claim by the [Employer] **OR** [Fund] **OR** [Acquirer] against the Sub-Contractor, or by any act or omission lose or prejudice the Sub-Contractor's right to make or proceed with such a claim against the insurers.

6.2   Any increased or additional premium required by insurers by reason of the Sub-Contractor's own claims record or other acts, omissions, matters or things particular to the Sub-Contractor shall be deemed to be within commercially reasonable rates.

6.3   The Sub-Contractor shall immediately inform the [Employer] **OR** [Fund] **OR** [Acquirer] if such insurance ceases to be available at commercially reasonable rates in order that the Sub-Contractor and the [Employer] **OR** [Fund] **OR** [Acquirer] can discuss means of best protecting the respective positions of the [Employer] **OR** [Fund] **OR** [Acquirer] and the Sub-Contractor in respect of the Development in the absence of such insurance.

6.4   The Sub-Contractor shall fully co-operate with any measures reasonably required by the [Employer] **OR** [Fund] **OR** [Acquirer], including (without limitation) completing any proposals for insurance and associated documents, maintaining such insurance at rates above commercially reasonable rates if the [Employer] **OR** [Fund] **OR** [Acquirer] undertakes in writing to reimburse the Sub-Contractor in respect of the net cost of such insurance to the Sub-Contractor above commercially reasonable rates or, if the [Employer] **OR** [Fund] **OR** [Acquirer] effects such insurance at rates at or above commercially reasonable rates, reimbursing the [Employer] **OR** [Fund] **OR** [Acquirer] in respect of what the net cost of such insurance to the [Employer] **OR** [Fund] **OR** [Acquirer] would have been at commercially reasonable rates.

6.5   As and when it is reasonably requested to do so by the [Employer] **OR** [Fund] **OR** [Acquirer] the Sub-Contractor shall produce for inspection documentary evidence (including, if required by the [Employer] **OR** [Fund] **OR** [Acquirer], the original of the relevant insurance documents) that its professional indemnity insurance is being maintained.

7.   This Agreement may be assigned by the [Employer] **OR** [Fund] **OR** [Acquirer] and its successors and assigns without the consent of the Sub-Contractor being required.

8.   Any notice to be given by the Sub-Contractor hereunder shall be deemed to be duly given if it is delivered by hand at or sent by registered post or recorded delivery to the above-mentioned address of the [Employer] **OR** [Fund] **OR** [Acquirer] or to the principal business address of the [Employer] **OR** [Fund] **OR** [Acquirer] for the time being, and any notice to be given by the [Employer] **OR** [Fund] **OR** [Acquirer] hereunder shall be deemed to be duly given if it is addressed to the Sub-Contractor and

delivered by hand at or sent by registered post or recorded delivery to the above-mentioned address of the Sub-Contractor or to the principal business address of the Sub-Contractor for the time being and, in the case of any such notices, the same shall if sent by registered post or recorded delivery be deemed to have been received forty-eight hours after being posted.

NOTE: The following clause will only be required in the case of Collateral Warranties in favour of the Employer or Fund.

9.✱   The Sub-Contractor shall within 7 (seven) working days of the [Employer's] **OR** [Fund's] request to do so, execute, in favour of any persons who have entered or shall enter into an agreement for the provision of finance in connection with the Development and/or in favour of any persons who have acquired or shall acquire any interest in or over the Development or any part thereof, a Deed in the form of this Deed, excluding this clause, or a similar form reasonably required by the [Employer] **OR** [Fund], and deliver the same to the [Employer] **OR** [Fund]; together in each case (if requested by the [Employer] **OR** [Fund]) with a guarantee (in form and substance reasonably required by the [Employer] **OR** [Fund]) from the ultimate parent company of the Sub-Contractor in respect of the Sub-Contractor's obligations pursuant to such Deed.

IN WITNESS whereof the Sub-Contractor has executed this Deed on the date first stated above.

COMMENTARY

See commentaries upon Forms 6, 7, Annexe 'E' to Form 15, 19 and 20. This Form is very much along the lines of Form 20. It may be appropriate in some cases for the Employer, Fund or Acquirer to be able to take over the employment of the Sub-Contractor, especially in the case of a Sub-Contractor undertaking design. If there is more than one beneficiary entitled to take over the Sub-Contractor's employment, an order of priority must be established. Also, it would usually be impracticable to take over the employment of one Sub-Contractor without taking over the employment of all the others.

Clause 1

This clause goes beyond a duty of care, as it contains a covenant actually to perform the Sub-Contract.

Clauses 5 and 6

These clauses may be irrelevant unless the Sub-Contractor has assumed design responsibility under the Sub-Contract.

Clause 9

This clause should help the Employer and the Fund in the event of the Contractor's insolvency.

FORM 22: CONSTRUCTION MANAGEMENT TRADE CONTRACT INCORPORATING JCT STANDARD FORM OF BUILDING CONTRACT, 1980 EDITION, PRIVATE WITH QUANTITIES, WITH OPTIONS FOR SECTIONAL COMPLETION

DATED   **199**

[   ]

('the Employer')

-and-

[   ]

('the Contractor')

CONSTRUCTION MANAGEMENT TRADE CONTRACT

for the construction of

[   ]

✽ An asterisk indicates that there is a relevant note in the commentary upon this Form.

THIS AGREEMENT made the   day of   One thousand nine hundred and ninety-

BETWEEN:

(1)

    [of] **OR** [whose registered office is at]

    ('the Employer'); and

(2)

    [of] **OR** [whose registered office is at]

    ('the Contractor');

INCORPORATES the Joint Contracts Tribunal Standard Form of Building Contract, 1980 Edition, Private With Quantities, as printed in   199   , incorporating Amendments 1 (January 1984), 2 (November 1986), 4 (July 1987), 5 (January 1988), 6 (July 1988), 7 (July 1988), 8 (April 1989), 9 (July 1990), 10 (March 1991) and 11 (July 1992, as corrected in September 1992), [as modified by the JCT Sectional Completion Supplement (1981 Edition, revised   199   )] and as amended by the Schedule.

IN WITNESS whereof the parties have executed this Deed in duplicate on the date first stated above.

SCHEDULE

Recitals

First recital✻

The Works are the construction of

forming part of a larger project briefly described as

('the Project') the balance of which (other than the Works) the Employer intends to execute himself or to have executed by persons employed or otherwise engaged by the Employer as referred to in clause 29.

The Drawings and Bills of Quantities were prepared by or under the direction of the [Architect] **OR** [Construction Manager].

Third recital

The Contract Drawings are numbered:

Fourth recital

Delete 'the Finance (No. 2) Act 1975' and substitute 'Part XIII (Miscellaneous Special Provisions) Chapter IV (Sub-Contractors in the Construction Industry) of the Income and Corporation Taxes Act 1988.'

Add the following Fifth recital:✱

'Fifth

the Employer and the Contractor have agreed a programme (hereinafter called 'the Programme') for the carrying out and completion of the Works.'

Articles of Agreement

Article 2: Contract Sum

The sum to be inserted is   pounds (£   ).

Article 3: [Architect] OR [Construction Manager]✱

The name and address of the [Architect] **OR** [Construction Manager] are

[of] **OR** [whose registered office is at]

Delete 'not being a person to whom the Contractor no later than 7 days after such nomination shall object for reasons considered to be sufficient by an Arbitrator appointed in accordance with article 5'.

[Add to Article 3:

'The expression 'Architect' shall be deemed to have been deleted throughout the Contract Documents, and 'Construction Manager' substituted.']

Article 4: Quantity Surveyor

The name and address of the Quantity Surveyor are

[of] **OR** [whose registered office is at]

Delete 'not being a person to whom the Contractor no later than 7 days after such nomination shall object for reasons considered to be sufficient by an Arbitrator appointed in accordance with article 5.'

Amendments to Conditions

Clause 1.3: Definitions

✷[The definition of 'Architect' shall be deleted, and the following substituted:

'Construction Manager: the person named in Article 3, or any successor duly appointed under Article 3, or otherwise agreed as the person to be the Construction Manager.']

The definition of 'Conditions' shall include any additional clauses or provisions hereby added, and the Conditions as hereby amended and added to.

✷The definition of 'Date of Possession' shall be deleted, and the following substituted:

'Date of Access: the date stated in the Appendix under the reference to clause 23.1: all references in the Contract Documents to the Date of Possession shall be deemed to be references to the Date of Access: and all references in the Contract Documents to possession of the site or Works by the Contractor shall be deemed to be references to such access as is described in clause 23.1.'

Add 'his successors and assigns' to the definition of 'Employer'.

The definition of 'Joint Names Policy' shall be deleted, and the following substituted:

'Joint Names Policy: a policy of insurance which includes the Contractor and the Employer and such other persons as the Employer may reasonably require (including, without limitation, any persons who have entered or shall enter into an agreement for the provision of finance in connection with the Project, any persons who have acquired or shall acquire any interest in or over the Project or any part thereof, and persons employed or otherwise engaged by the Employer as referred to in clause 29) as the insured.'

[The definitions of 'Nominated Sub-Contract', 'Nominated Sub-Contractor', and 'Numbered Documents' shall be deleted.]

The following definitions shall be added:

'Programme: see Fifth recital: the Programme shall be deemed to form part of the Contract Bills.

Project: see First recital.'

Clause 2: Contractor's obligations✻

Delete clause 2.1, and substitute:

'Subject to receipt of notice from the [Architect] **OR** [Construction Manager] to commence works on site, the Contractor shall upon and subject to the Conditions carry out and complete the Works in compliance with the Contract Documents, and reasonably in accordance with the progress of the Project, and in conformity with all reasonable instructions of the [Architect] **OR** [Construction Manager] (including, without limitation, instructions for the variation of the Programme), using materials and workmanship of the quality and standards specified in the Contract Documents, provided that where and to the extent that approval of the quality of materials or of the standards of workmanship is a matter for the opinion of the [Architect] **OR** [Construction Manager], such quality and standards shall be to the reasonable satisfaction of the [Architect] **OR** [Construction Manager]. If the regular progress of the Project (including, without limitation, any part thereof being or to be executed by the Employer himself or by persons employed or otherwise engaged by the Employer as referred to in clause 29) is materially adversely affected by any act, omission or default of the Contractor, his servants or agents, or any sub-contractor employed by the Contractor, the Employer shall within a reasonable time of such material adverse effect becoming apparent give written notice thereof to the Contractor, and any loss and/or expense thereby caused to the Employer (whether suffered or incurred by the Employer himself or by persons employed or otherwise engaged by the Employer as referred to in clause 29) may be deducted from any monies due or to become due to the Contractor or may be recoverable from the Contractor as a debt. The notice to commence works on site may be issued within [   ] days of the date of formation of the Contract, and shall specify a Date of Access not less than [   ] days and not more than [   ] days after the date of issue of such notice. It shall be within the absolute discretion of the Employer whether or not notice to commence works on site shall be given. Time shall be of the essence with regard to the time limit for the issue of the notice to commence works on site.'

Clause 2.2.1 shall be deleted.

Clause 13: Variations and provisional sums

Delete clause 13.1.3, and substitute:

'.3   compliance with the [Architect's] **OR** [Construction Manager's] instructions under clause 2.1 or clause 29.1 [and nomination of a Sub-Contractor to supply and fix materials or goods or to execute work of which the measured quantities have been set out and priced by the Contractor in the Contract Bills for supply and fixing or executing by the Contractor.]'

Clause 18: Partial possession by Employer

In clause 18.1, delete 'and the consent of the Contractor (which consent shall not be unreasonably withheld) has been obtained....'.

Clauses 19.1.1 and 19.1.2: Assignment

Delete clause 19.1.1 and the first sentence of clause 19.1.2, and substitute:

'19.1.1 The Contractor shall not, without the written consent of the Employer, assign this Contract.

19.1.2 Where clause 19.1.2 is stated in the Appendix to apply, the Employer may assign this Contract.'

[In clause 19.2, delete 'other than a Nominated Sub-Contractor', and add:

'All references in the Contract Documents to Nominated Sub-Contractors shall be disregarded'.]

[Delete clause 19.5.1, and substitute:

'Save as otherwise expressed in the Conditions, the Contractor shall remain wholly responsible for carrying out and completing the Works in all respects in accordance with clause 2.1, notwithstanding the sub-contracting of the supply or fixing of any materials or goods, or of the execution of any work'.]

[Clause 19.5.2 shall be deleted.]

Clause 22A: Erection of new buildings – All Risks Insurance of the Works by the Contractor✻

Clause 22A shall be deleted.

Clause 22B: Erection of new buildings – All Risks Insurance of the Works by the Employer✻

In clause 22B.1, delete 'Works', and substitute 'Project'.

Clause 22C: Insurance of existing structures – Insurance of Works in or extensions to existing structures✻

In clause 22C.2, delete 'Works', and substitute 'Project'.

Clause 23: Date of Possession, completion and postponement

Clause 23.1.1 shall be deleted, and the following substituted:

'On the Date of Access, sufficient access to the site shall be given to the Contractor, who shall thereupon begin the Works, regularly and diligently proceed with the same and shall complete the same on or before the Completion Date'.

In clause 23.1.2, delete 'possession', and substitute 'access'.

In clause 23.3.1, delete 'and, subject to clause 18, the Employer shall not be entitled to take possession of any part or parts of the Works until that date'.

Clause 23.3.2 shall be deleted, and the following substituted:

'The Employer may use or occupy the site or the Works or part thereof whether for the purposes of the execution of the balance of the Project (other than the Works), storage of his goods or for any other purpose whatsoever before the date of issue of the certificate of Practical Completion by the [Architect] **OR** [Construction Manager]. Before such use or occupation, the Employer shall notify the insurers under clause 22B or 22C.2 and .4, whichever may be applicable, and obtain confirmation that such use or occupation will not prejudice the insurance'.

Clause 23.3.3 shall be deleted.

Clause 24: Damages for non-completion✳

[In clause 24.2.1, delete 'as the Employer may require in writing not later than the date of the Final Certificate'.]

OR

[Delete clause 24.2.1, and substitute:

'Subject to the issue of a certificate under clause 24.1, the Contractor shall pay or allow to the Employer a sum equivalent to any loss or damage suffered or incurred by the Employer and caused by the failure of the Contractor as aforesaid, and the Employer may set-off and deduct the same from any monies due or to become due to the Contractor under this Contract (including any balance stated as due to the Contractor in the Final Certificate) or the Employer may recover the same from the Contractor as a debt'.]

Clause 25.4: Relevant Events

[Clause 25.4.7 shall be deleted.]

Clause 25.4.10 shall be deleted.

Add to clause 25.4 as an additional sub-clause:

'any breach of this Contract by the Employer or any act or omission on the part of the Employer, the [Architect] **OR** [Construction Manager] or the Quantity Surveyor'.

Clause 27: Determination by Employer

Clause 27.2.4 shall be deleted.

The last sentence of clause 27.5.3 shall be deleted.

The proviso to clause 27.6.4.1 shall be deleted.

[Clause 27.7 shall be deleted.]

OR

[In clause 27.7.1, delete '6 months', and substitute '12 (twelve) months'.

In clause 27.7.2, delete '6 month', and substitute '12 (twelve) month'.]

Clause 28A: Determination by Employer or Contractor

Clause 28A.5.5 shall be deleted.

Clause 29: Works by Employer or persons employed or engaged by Employer

Clause 29.1 shall be deleted, and the following substituted:

'In regard to any work forming part of the Project, but not forming part of this Contract, and which is to be carried out by the Employer himself or by persons employed or otherwise engaged by him, the Contractor shall permit, and co-operate with and in, the execution of such work, and shall act in conformity with all reasonable instructions of the [Architect] **OR** [Contract Administrator] relating to such work, its execution and the Contractor's required co-operation'.

In clause 29.2, delete 'Where the Contract Bills do not provide the information referred to in clause 29.1...' and substitute 'Where the Employer requires the execution of work not forming part of the Project and not forming part of this Contract...'.

Clause 30: Certificates and Payments

In clause 30.1.1.2, delete:

'Notwithstanding the fiduciary interest of the Employer in the Retention as stated in clause 30.5.1...'.

Clause 30.5.1 shall be deleted, and the following substituted:

'30.5.1   the Employer's interest in the Retention shall not be fiduciary, either as trustee for the Contractor [or any Nominated Sub-Contractor] or any other person, or in any other capacity; the relationship of the Employer and the Contractor with regard to the Retention shall be solely that of debtor and unsecured creditor, subject to the terms hereof; and the Employer shall have no obligation to invest the Retention or any part thereof;'

Clause 30.5.3 shall be deleted, and the following substituted:

'30.5.3   the Employer shall have no obligation to segregate the Retention or any part thereof in a separate banking account, or in any other manner whatsoever; and shall be entitled to the full beneficial interest in the Retention and every part thereof (and, without limitation, interest thereon and income arising therefrom) unless and until the Retention is paid to the Contractor pursuant to this Contract.'

Clause 30.3 shall be deleted.

In clause 30.9.2, the last word ('either') of the first paragraph shall be deleted.

In clause 30.9.2.1, the last word ('or') shall be deleted.

Clause 30.9.2.2 shall be deleted.

The concluding words of clause 30.9.2 ('whichever shall be the earlier') shall be deleted.

Clause 31: Finance (No.2) Act 1975 – statutory tax deduction scheme

References to 'the Act', 'the Regulations' or any legislation shall be interpreted to include references to any amendment or re-enactment thereof.

[**Clause 35: Nominated Sub-Contractors✻**

Clause 35.18.2 shall be deleted.

In clause 35.21, the word 'not' shall be deleted where it occurs for the second time.

Clause 35.24.9 shall be deleted, and the following substituted:

'35.24.9 The amount properly payable to the Nominated Sub-Contractor under the Sub-Contract resulting from such further nomination shall be included in the amount stated as due in Interim Certificates and added to the Contract Sum. Provided that any extra amount, payable by the Employer in respect of the Sub-Contractor nominated under the further nomination over the price of the Nominated Sub-Contractor who first entered into a sub-contract in respect of the relevant sub-contract works, resulting from such further nomination, may at the time or at any time after such amount is certified in respect of the Sub-Contractor nominated under the further nomination, be deducted by the Employer from monies due or to become due to the Contractor under this Contract, or may be recoverable from the Contractor by the Employer as a debt.']

OR

[Clause 35 shall be deleted.]

Clause 38.4: Provisions relating to clause 38

Clause 38.4.8.1 shall be deleted.

Clause 39.5: Provisions relating to clause 39

Clause 39.5.8.1 shall be deleted.

Clause 40: Use of price adjustment formulae

Clause 40.7.2.1 shall be deleted.

Clause 41: Settlement of Disputes – Arbitration

In clause 41.2.1, add after the opening proviso:

'the Employer and any person or persons employed or otherwise engaged by the Employer as referred to in clause 29, or...'.

Additional Conditions

The following clauses shall be added:

'42:   Collateral Warranties∗

42.1   The Contractor shall within 7 (seven) working days of the Employer's request so to do:

42.1.1   execute, in favour of any persons who have entered or shall enter into an agreement for the provision of finance in connection with the Project, a Deed in the form annexed as Annexe 'A', or a similar form reasonably required by the Employer, and deliver the same to the Employer, together with a guarantee in the form annexed as Annexe 'C', or a similar form reasonably required by the Employer, from the ultimate parent company of the Contractor, namely [   ] [Limited] **OR** [PLC], in respect of the Contractor's obligations pursuant to such Deed; and

42.1.2   execute, in favour of any persons who have acquired or shall acquire any interest in or over the Project or any part thereof, a Deed in the form annexed as Annexe 'B', or a similar form reasonably required by the Employer, and deliver the same to the Employer, together with a guarantee in the form annexed as Annexe 'C', or a similar form reasonably required by the Employer, from the ultimate parent company of the Contractor, namely [   ] [Limited] **OR** [PLC], in respect of the Contractor's obligations pursuant to such Deed.

42.2   The Contractor shall (without request) use reasonable endeavours to procure that each sub-contract or supply contract with each Nominated [Sub-Contractor or] Supplier, and any other sub-contractor or supplier (if requested by the Employer), shall contain obligations on the relevant sub-contractor or supplier to execute, in favour of the Employer and/or (within 7 (seven) working days of the Employer's request so to do) any persons who have entered or shall enter into an agreement for the provision of finance in connection with the Project and/or in favour of any persons who have acquired or shall acquire any interest in or over the Project or any part thereof, a Deed in the form annexed as Annexe 'D', or a similar form reasonably required by the Employer, and deliver the same to the Employer; together in each case (if requested by the Employer) with a guarantee in the form annexed as Annexe 'E', or a similar form reasonably required by the Employer, from the ultimate parent company of the relevant sub-contractor or supplier in respect of the sub-contractor's or supplier's obligations pursuant to such Deed. The Contractor shall enforce such obligations, or such modified obligations as are referred to below.

42.3   If, despite the Contractor having used such reasonable endeavours, the sub-contractor or supplier will not accept such obligations, or will only accept them in a modified form, the Contractor shall notify the Employer, who may agree in writing that the relevant sub-contract or supply contract need not contain such obligations, or that the relevant obligations may be in a modified form agreeable to the sub-contractor or supplier.

42.4   Failing such agreement by the Employer in the case of a proposed Nominated [Sub-Contractor or] Supplier, the Contractor shall not be obliged to enter into a sub-contract or supply contract with that Nominated [Sub-Contractor or] Supplier.

42.5   Failing such agreement by the Employer in the case of any other sub-contractor or supplier, the Contractor shall not enter into a relevant sub-contract or supply contract with that sub-contractor or supplier.

42.6   The above obligations for the provision of Deeds and guarantees in favour of third parties shall continue notwithstanding termination of this Contract, or determination of the Contractor's employment hereunder, in either case for any reason whatsoever, including (without limitation) breach by the Employer. However, any such Deed given after such termination or determination shall be amended by the Employer so as to refer to the fact and date of such termination or determination, to omit any obligation to continue to exercise skill, care or diligence or fulfil obligations pursuant to this Contract after such termination or determination, and to omit any provision enabling a third party to assume the position of employer of the Contractor.

43:   **[Performance Bond] OR [Parent Company Guarantee]**

Upon execution of this Contract, the Contractor shall deliver to the Employer a performance bond from [   ] [Limited] **OR** [PLC] in an amount of   % (per cent) of the Contract Sum in the form annexed as Annexe 'F' **OR** a parent company guarantee from its ultimate parent company, namely [   ] [Limited] **OR** [PLC], for its performance of this Contract in the form annexed as Annexe 'F'.

44:   **Proscribed Materials**

44.1   The Contractor warrants:

44.1.1　that he has not used or specified and will not use or specify for use;

44.1.2　that he has exercised and will continue to exercise reasonable skill, care and diligence to see that there are not used;

44.1.3　that he is not aware and has no reason to suspect or believe that there have been or will be used; and

44.1.4　that he will promptly notify the Employer in writing if he becomes aware or has reason to suspect or believe that there have been or will be used;

in or in connection with the Project, any of the materials or substances identified in clause 44.2.

44.2　The said materials or substances are:

44.2.1　high alumina cement in structural elements;

44.2.2　wood wool slabs in permanent formwork to concrete;

44.2.3　calcium chloride in admixtures for use in reinforced concrete;

44.2.4　asbestos products;

44.2.5　naturally occurring aggregates for use in reinforced concrete which do not comply with British Standard 882: 1983 and/or naturally occurring aggregates for use in concrete which do not comply with British Standard 8110: 1985.'

APPENDIX*

COMMENTARY

General

See commentary upon Form 12, and Chapter 5. The further comments given below relate to amendments and items not found in Form 12.

First recital

A distinction is drawn between the 'Works' to be executed under this Trade Contract and the 'Project', which will be the aggregate of the 'Works' executed under all the Trade Contracts.

Fifth recital

A pre-contract agreed Programme is required, as in the JCT NSC/T procedure, in order to assist in co-ordinating the efforts of all the Trade Contractors.

Article 3 and clause 1.3: Definition of Architect

The term 'Architect' may be used in this Form if the person acting as construction manager is entitled under statute to use the name 'Architect'. Otherwise, the term 'Construction Manager' should be used. The Form contains alternatives accordingly.

Clause 1.3: Date of Access

'Date of Access' is substituted for 'Date of Possession', as several Trade Contractors will be on site concurrently.

Clause 2.1: Contractor's obligations

Clause 2.1 now provides for a notice to commence, to be issued within a limited time after entry into the Contract, which must specify a Date of Access. The power to issue instructions is much expanded, in order to give the degree of control required in a construction management Project, with multiple main Contractors concurrently on site. Cf. clause 2.1 of JCT Form NSC/C. A provision based upon clause 4.39 of NSC/C is also included. See also the amendments to clauses 13 and 29.

Clause 22: Insurance of the Works

The nature of construction management renders it convenient for the Employer to insure the Project under clause 22B or 22C, rather than for the Contractor to insure the Project.

Clause 24: Damages for non-completion

A choice is given of liquidated and ascertained damages, or of damages at large.

Clause 35: Nominated Sub-Contractors

The Form contains optional amendments in order to delete all references to Nominated Sub-Contractors, because they may well be inappropriate under a Construction Management Trade Contract.

Clause 42: Collateral Warranties

For Annexes 'A', 'B', 'C', 'D' and 'E', see Forms 25, 26, 28, 27 and 29.

Appendix

The published Appendix should be completed and attached. See the equivalent commentary upon Form 12. Suggested Appendix entries regarding Defects Liability Period, clause 21.2.1 insurance, Date of Access, Deferment of the Date of Access and Liquidated and Ascertained Damages (**if damages for delay are to be at large**) are as follows:

'Defects Liability Period	17.2	The period of [24 (twenty-four)] months from the day named in the Certificate of Practical Completion of the Works, or the period from the day named in the Certificate of Practical Completion of the Works to the expiry of [12 (twelve)] months from the practical completion of the Project, whichever is the [shorter] **OR** [longer] period.
Insurance – liability of Employer	21.2.1	Insurance is not required.
Date of Access	23.1.1	The Date of Access specified in the notice to commence works on site issued pursuant to clause 2.1.
Deferment of the Date of Access	23.1.2 25.4.13 26.1	Clause 23.1.2 [applies] **OR** [does not apply] Period of deferment if it is to be less than 6 (six) weeks is: [ ]
Liquidated and ascertained damages	24.2	Not applicable. Damages for delay shall be at large'

The above items will need to be suitably adapted if there is to be sectional completion.

FORM 23: CONSTRUCTION MANAGEMENT TRADE CONTRACT INCORPORATING JCT STANDARD FORM OF BUILDING CONTRACT, 1980 EDITION, PRIVATE WITH APPROXIMATE QUANTITIES, WITH OPTIONS FOR SECTIONAL COMPLETION

DATED _____ **199**

[   ]

('the Employer')

-and-

[   ]

('the Contractor')

**CONSTRUCTION MANAGEMENT
TRADE CONTRACT**

for the construction of

[   ]

✳ An asterisk indicates that there is a relevant note in the commentary upon this Form.

THIS AGREEMENT made the   day of   One thousand nine hundred and ninety-

BETWEEN:

(1)

       [of] **OR** [whose registered office is at]

       ('the Employer'); and

(2)

       [of] **OR** [whose registered office is at]

       ('the Contractor');

INCORPORATES the Joint Contracts Tribunal Standard Form of Building Contract, 1980 Edition, Private With Approximate Quantities, as printed in   199 , incorporating Amendments 1 (January 1984), 2 (November 1986), 4 (July 1987), 5 (January 1988), 6 (July 1988), 7 (July 1988), 8 (April 1989), 9 (July 1990), 10 (March 1991) and 11 (July 1992, as corrected in September 1992), [as modified by the JCT Sectional Completion Supplement (1981 Edition, revised   199)], and as amended by the Schedule.

IN WITNESS whereof the parties have executed this Deed in duplicate on the date first stated above.

SCHEDULE

Recitals

First recital

The Works are the construction of

forming part of a larger project briefly described as

('the Project') the balance of which (other than the Works) the Employer intends to execute himself or to have executed by persons employed or otherwise engaged by the Employer as referred to in clause 29.

The Drawings and Bills of Approximate Quantities were prepared by or under the direction of the [Architect] **OR** [Construction Manager].

Second recital

The Tender Price is   pounds (£   ).

Third recital

The Contract Drawings are numbered:

Fourth recital

Delete 'the Finance (No. 2) Act 1975' and substitute 'Part XIII (Miscellaneous Special Provisions) Chapter IV (Sub-Contractors in the Construction Industry) of the Income and Corporation Taxes Act 1988.'

Add the following Fifth recital:

'Fifth

the Employer and the Contractor have agreed a programme (hereinafter called 'the Programme') for the carrying out and completion of the Works.'

Articles of Agreement

Article 3: [Architect] OR [Construction Manager]

The name and address of the [Architect] **OR** [Construction Manager] are

[of] **OR** [whose registered office is at]

Delete 'not being a person to whom the Contractor no later than 7 days after such nomination shall object for reasons considered to be sufficient by an Arbitrator appointed in accordance with article 5'.

[Add to Article 3:

'The expression 'Architect' shall be deemed to have been deleted throughout the Contract Documents, and 'Construction Manager' substituted.']

Article 4: Quantity Surveyor

The name and address of the Quantity Surveyor are

[of] **OR** [whose registered office is at]

Delete 'not being a person to whom the Contractor no later than 7 days after such nomination shall object for reasons considered to be sufficient by an Arbitrator appointed in accordance with article 5.'

Amendments to Conditions

Clause 1.3: Definitions

[The definition of 'Architect' shall be deleted, and the following substituted:

'Construction Manager: the person named in Article 3, or any successor duly appointed under Article 3, or otherwise agreed as the person to be the Construction Manager.']

The definition of 'Conditions' shall include any additional clauses or provisions hereby added, and the Conditions as hereby amended and added to.

The definition of 'Date of Possession' shall be deleted, and the following substituted:

'Date of Access: the date stated in the Appendix under the reference to clause 23.1: all references in the Contract Documents to the Date of Possession shall be deemed to be references to the Date of Access: and all references in the Contract Documents to possession of the site or Works by the Contractor shall be deemed to be references to such access as is described in clause 23.1.'

Add 'his successors and assigns' to the definition of 'Employer'.

The definition of 'Joint Names Policy' shall be deleted, and the following substituted:

'Joint Names Policy: a policy of insurance which includes the Contractor and the Employer and such other persons as the Employer may reasonably require (including, without limitation, any persons who have entered or shall enter into an agreement for the provision of finance in connection with the Project, any persons who have acquired or shall acquire any interest in or over the Project or any part thereof, and persons employed or otherwise engaged by the Employer as referred to in clause 29) as the insured.'

[The definitions of 'Nominated Sub-Contract', 'Nominated Sub-Contractor', and 'Numbered Documents' shall be deleted.]

The following definitions shall be added:

'Programme: see Fifth recital: the Programme shall be deemed to form part of the Contract Bills.

Project: see First recital.'

Clause 2: Contractor's obligations

Delete clause 2.1, and substitute:

'Subject to receipt of notice from the [Architect] **OR** [Construction Manager] to commence works on site, the Contractor shall upon and subject to the Conditions carry out and complete the Works in compliance with the Contract Documents, and reasonably in accordance with the progress of the Project, and in conformity with all reasonable instructions of the [Architect] **OR** [Construction Manager] (including, without limitation, instructions for the variation of the Programme), using materials and workmanship of the quality and standards specified in the Contract Documents, provided that where and to the extent that approval of the quality of materials or of the standards of workmanship is a matter for the opinion of the [Architect] **OR** [Construction Manager], such quality and standards shall be to the reasonable satisfaction of the [Architect] **OR** [Construction Manager]. If the regular progress of the Project (including, without limitation, any part thereof being or to be executed by the Employer himself or by persons employed or otherwise engaged by the Employer as referred to in clause 29) is materially adversely affected by any act, omission or default of the Contractor, his servants or agents, or any sub-contractor employed by the Contractor, the Employer shall within a reasonable time of such material adverse effect becoming apparent give written notice thereof to the Contractor, and any loss and/or expense thereby caused to the Employer (whether suffered or incurred by the Employer himself or by persons employed or otherwise engaged by the Employer as referred to in clause 29) may be deducted from any monies due or to become due to the Contractor or may be recoverable from the Contractor as a debt. The notice to commence works on site may be issued within [   ] days of the date of formation of the Contract, and shall specify a Date of Access not less than [   ] days and not more than [   ] days after the date of issue of such notice. It shall be within the absolute discretion of the Employer whether or not notice to commence works on site shall be given. Time shall be of the essence with regard to the time limit for the issue of the notice to commence works on site.'

Clause 2.2.1 shall be deleted.

Clause 14: Measurement and Valuation of Work including variations and provisional sums

Delete clause 14.1.3, and substitute:

'.3   compliance with the [Architect's] **OR** [Construction Manager's] instructions under clause 2.1 or clause 29.1 [and nomination of a Sub-Contractor to supply and fix materials or goods or to execute work of which the measured quantities have been set out and priced by the Contractor in the Contract Bills for supply and fixing or execution by the Contractor].'

Clause 18: Partial possession by Employer

In clause 18.1, delete 'and the consent of the Contractor (which consent shall not be unreasonably withheld) has been obtained....'

Clauses 19.1.1 and 19.1.2: Assignment

Delete clause 19.1.1 and the first sentence of clause 19.1.2, and substitute:

'19.1.1 The Contractor shall not, without the written consent of the Employer, assign this Contract.

19.1.2 Where clause 19.1.2 is stated in the Appendix to apply, the Employer may assign this Contract.'

[In clause 19.2, delete 'other than a Nominated Sub-Contractor', and add:

'All references in the Contract Documents to Nominated Sub-Contractors shall be disregarded'.]

[Delete clause 19.5.1, and substitute:

'Save as otherwise expressed in the Conditions, the Contractor shall remain wholly responsible for carrying out and completing the Works in all respects in accordance with clause 2.1, notwithstanding the sub-contracting of the supply or fixing of any materials or goods, or of the execution of any work'.]

[Clause 19.5.2 shall be deleted.]

Clause 22A: Erection of new buildings – All Risks Insurance of the Works by the Contractor

Clause 22A shall be deleted.

Clause 22B: Erection of new buildings – All Risks Insurance of the Works by the Employer

In clause 22B.1, delete 'Works', and substitute 'Project'.

Clause 22C: Insurance of existing structures – Insurance of Works in or extensions to existing structures

In clause 22C.2, delete 'Works', and substitute 'Project'.

Clause 23: Date of Possession, completion and postponement

Clause 23.1.1 shall be deleted, and the following substituted:

'On the Date of Access, sufficient access to the site shall be given to the Contractor, who shall thereupon begin the Works, regularly and diligently proceed with the same and shall complete the same on or before the Completion Date'.

In clause 23.1.2, delete 'possession', and substitute 'access'.

In clause 23.3.1, delete 'and, subject to clause 18, the Employer shall not be entitled to take possession of any part or parts of the Works until that date'.

Clause 23.3.2 shall be deleted, and the following substituted:

'The Employer may use or occupy the site or the Works or part thereof whether for the purposes of the execution of the balance of the Project (other than the Works), storage of his goods or for any other purpose whatsoever before the date of issue of the certificate of Practical Completion by the [Architect] **OR** [Construction Manager]. Before such use or occupation, the Employer shall notify the insurers under clause 22B or 22C.2 and .4, whichever may be applicable, and obtain confirmation that such use or occupation will not prejudice the insurance'.

Clause 23.3.3 shall be deleted.

Clause 24: Damages for non-completion

[In clause 24.2.1, delete 'as the Employer may require in writing not later than the date of the Final Certificate'.]

OR

[Delete clause 24.2.1, and substitute:

'Subject to the issue of a certificate under clause 24.1, the Contractor shall pay or allow to the Employer a sum equivalent to any loss or damage suffered or incurred by the Employer and caused by the failure of the Contractor as aforesaid, and the Employer may set-off and deduct the same from any monies due or to become due to the Contractor under this Contract (including any balance stated as due to the Contractor in the Final Certificate) or the Employer may recover the same from the Contractor as a debt'.]

Clause 25.4: Relevant Events

[Clause 25.4.7 shall be deleted.]

Clause 25.4.10 shall be deleted.

Add to clause 25.4 as an additional sub-clause:

'any breach of this Contract by the Employer or any act or omission on the part of the Employer, the [Architect] **OR** [Construction Manager] or the Quantity Surveyor'.

Clause 27: Determination by Employer

Clause 27.2.4 shall be deleted.

The last sentence of clause 27.5.3 shall be deleted.

The proviso to clause 27.6.4.1 shall be deleted.

[Clause 27.7 shall be deleted.]

OR

[In clause 27.7.1, delete '6 months', and substitute '12 (twelve) months'.

In clause 27.7.2, delete '6 month', and substitute '12 (twelve) month'.]

Clause 28A: Determination by Employer or Contractor

Clause 28A.5.5 shall be deleted.

Clause 29: Works by Employer or persons employed or engaged by Employer

Clause 29.1 shall be deleted, and the following substituted:

'In regard to any work forming part of the Project, but not forming part of this Contract, and which is to be carried out by the Employer himself or by persons employed or otherwise engaged by him, the Contractor shall permit, and co-operate with and in, the execution of such work, and shall act in conformity with all reasonable instructions of the [Architect] **OR** [Contract Administrator] relating to such work, its execution and the Contractor's required co-operation'.

In clause 29.2, delete 'Where the Contract Bills do not provide the information referred to in clause 29.1...' and substitute 'Where the Employer requires the execution of work not forming part of the Project and not forming part of this Contract...'.

Clause 30: Certificates and Payments

In clause 30.1.1.2, delete:

'Notwithstanding the fiduciary interest of the Employer in the Retention as stated in clause 30.5.1...'.

Clause 30.5.1 shall be deleted, and the following substituted:

'30.5.1   the Employer's interest in the Retention shall not be fiduciary, either as trustee for the Contractor [or any Nominated Sub-Contractor] or any other person, or in any other capacity; the relationship of the Employer and the Contractor with regard to the Retention shall be solely that of debtor and unsecured creditor, subject to the terms hereof; and the Employer shall have no obligation to invest the Retention or any part thereof;'

Clause 30.5.3 shall be deleted, and the following substituted:

'30.5.3   the Employer shall have no obligation to segregate the Retention or any part thereof in a separate banking account, or in any other manner whatsoever; and shall be entitled to the full beneficial interest in the Retention and every part thereof (and, without limitation, interest thereon and income arising therefrom) unless and until the Retention is paid to the Contractor pursuant to this Contract.'

Clause 30.3 shall be deleted.

In clause 30.9.2, the last word ('either') of the first paragraph shall be deleted.

In clause 30.9.2.1, the last word ('or') shall be deleted.

Clause 30.9.2.2 shall be deleted.

The concluding words of clause 30.9.2 ('whichever shall be the earlier') shall be deleted.

Clause 31: Finance (No. 2) Act 1975 – statutory tax deduction scheme

References to 'the Act', 'the Regulations' or any legislation shall be interpreted to include references to any amendment or re-enactment thereof.

[Clause 35: Nominated Sub-Contractors

Clause 35.18.2 shall be deleted.

In clause 35.21, the word 'not' shall be deleted where it occurs for the second time.

Clause 35.24.9 shall be deleted, and the following substituted:

'35.24.9 The amount properly payable to the Nominated Sub-Contractor under the Sub-Contract resulting from such further nomination shall be included in the amount stated as due in Interim Certificates and added to the Contract Sum. Provided that any extra amount, payable by the Employer in respect of the Sub-Contractor nominated under the further nomination over the price of the Nominated Sub-Contractor who first entered into a sub-contract in respect of the relevant sub-contract works, resulting from such further nomination, may at the time or at any time after such amount is certified in respect of the Sub-Contractor nominated under the further nomination, be deducted by the Employer from monies due or to become due to the Contractor under this Contract, or may be recoverable from the Contractor by the Employer as a debt.']

OR

[Clause 35 shall be deleted.]

Clause 38.4: Provisions relating to clause 38

Clause 38.4.8.1 shall be deleted.

Clause 39.5: Provisions relating to clause 39

Clause 39.5.8.1 shall be deleted.

Clause 40: Use of price adjustment formulae

Clause 40.7.2.1 shall be deleted.

Clause 41: Settlement of Disputes – Arbitration

In clause 41.2.1, add after the opening proviso:

'the Employer and any person or persons employed or otherwise engaged by the Employer as referred to in clause 29, or...'.

Additional Conditions✳

COMMENTARY

Continue as in Form 22, and see the commentary upon that Form.

However, clause 43 should refer to 'Tender Price', rather than 'Contract Sum'.

FORM 24: CONSTRUCTION MANAGEMENT TRADE CONTRACT INCORPORATING JCT STANDARD FORM OF BUILDING CONTRACT, 1980 EDITION, PRIVATE WITHOUT QUANTITIES

DATED _____ **199**

[   ]

('the Employer')

-and-

[   ]

('the Contractor')

CONSTRUCTION MANAGEMENT TRADE CONTRACT

for the construction of

[   ]

✱ **An asterisk indicates that there is a relevant note in the commentary upon this Form.**

THIS AGREEMENT made the   day of   One thousand nine hundred and ninety-

BETWEEN:

(1)

[of] **OR** [whose registered office is at]

('the Employer'); and

(2)

[of] **OR** [whose registered office is at]

('the Contractor');

INCORPORATES the Joint Contracts Tribunal Standard Form of Building Contract, 1980 Edition, Private Without Quantities, as printed in   199   , incorporating Amendments 1 (January 1984), 2 (November 1986), 3 (March 1987), 4 (July 1987), 5 (January 1988), 6 (July 1988), 8 (April 1989), 9 (July 1990), 10 (March 1991), and 11 (July 1992, as corrected in September 1992), and as amended by the Schedule.

IN WITNESS whereof the parties have executed this Deed in duplicate on the date first stated above.

SCHEDULE

Recitals

First recital

The Works are the construction of

forming part of a larger project briefly described as

('the Project') the balance of which (other than the Works) the Employer intends to execute himself or to have executed by persons employed or otherwise engaged by the Employer as referred to in clause 29.

The Contract Drawings are numbered:

The Contract Drawings and the [Specification] **OR** [Schedules of Work] were prepared by or under the direction of [the Architect] **OR** [Construction Manager].

Second recital

[Alternative A shall apply.]

OR

[Alternative B shall apply. The Contractor has supplied the Employer with a [Contract Sum Analysis as defined in Clause 1.3] **OR** [Schedule of Rates on which the Contract Sum is based].]

Third recital

Delete 'the Finance (No. 2) Act 1975' and substitute 'Part XIII (Miscellaneous Special Provisions) Chapter IV (Sub-Contractors in the Construction Industry) of the Income and Corporation Taxes Act 1988.'

Add the following Fourth recital:

'Fourth

the Employer and the Contractor have agreed a programme (hereinafter called 'the Programme') for the carrying out and completion of the Works.'

Articles of Agreement

Article 2: Contract Sum

The sum to be inserted is   pounds (£   ).

Article 3: [Architect] OR [Construction Manager]

The name and address of the [Architect] **OR** [Construction Manager] are

[of] **OR** [whose registered office is at]

Delete 'not being a person to whom the Contractor no later than 7 days after such nomination shall object for reasons considered to be sufficient by an Arbitrator appointed in accordance with article 5'.

[Add to Article 3:

'The expression 'Architect' shall be deemed to have been deleted throughout the Contract Documents, and 'Construction Manager' substituted.]

[Article 4A: Quantity Surveyor

The name and address of the Quantity Surveyor are

[of] **OR** [whose registered office is at]

Delete 'not being a person to whom the Contractor no later than 7 days after such nomination shall object for reasons considered to be sufficient by an Arbitrator appointed in accordance with article 5.']

OR

[Article 4B: Exercise of functions of Quantity Surveyor

The functions ascribed by the conditions to the 'Quantity Surveyor' shall be exercised by

[of] **OR** [whose registered office is at]

Delete 'not being a person to whom the Contractor no later than 7 days after such nomination shall object for reasons considered to be sufficient by an Arbitrator appointed in accordance with article 5.']

Amendments to Conditions

Clause 1.3: Definitions

[The definition of 'Architect' shall be deleted, and the following substituted:

'Construction Manager: the person named in Article 3, or any successor duly appointed under Article 3, or otherwise agreed as the person to be the Construction Manager.']

The definition of 'Conditions' shall include any additional clauses or provisions hereby added, and the Conditions as hereby amended and added to.

The definition of 'Date of Possession' shall be deleted, and the following substituted:

'Date of Access: the date stated in the Appendix under the reference to clause 23.1: all references in the Contract Documents to the Date of Possession shall be deemed to be references to the Date of Access: and all references in the Contract Documents to possession of the site or Works by the Contractor shall be deemed to be references to such access as is described in clause 23.1.'

Add 'his successors and assigns' to the definition of 'Employer'.

The definition of 'Joint Names Policy' shall be deleted, and the following substituted:

'Joint Names Policy: a policy of insurance which includes the Contractor and the Employer and such other persons as the Employer may reasonably require (including, without limitation, any persons who have entered or shall enter into an agreement for the provision of finance in connection with the Project, any persons who have acquired or shall acquire any interest in or over the Project or any part thereof, and persons employed or otherwise engaged by the Employer as referred to in clause 29) as the insured.'

[The definitions of 'Nominated Sub-Contract', 'Nominated Sub-Contractor', and 'Numbered Documents' shall be deleted.]

The following definitions shall be added:

'Programme: see Fourth recital: the Programme shall be deemed to form part of the [Specification] **OR** [Schedules of Work].

Project: see First recital.'

Clause 2: Contractor's obligations

Delete clause 2.1, and substitute:

'Subject to receipt of notice from the [Architect] **OR** [Construction Manager] to commence works on site, the Contractor shall upon and subject to the Conditions carry out and complete the Works in compliance with the Contract Documents, and reasonably in accordance with the progress of the Project, and in conformity with all reasonable instructions of the [Architect] **OR** [Construction Manager] (including, without limitation, instructions for the variation of the Programme), using materials and workmanship of the quality and standards specified in the Contract Documents, provided that where and to the extent that approval of the quality of materials or of the standards of workmanship is a matter for the opinion of the [Architect] **OR** [Construction Manager], such quality and standards shall be to the reasonable satisfaction of the [Architect] **OR** [Construction Manager]. If the regular progress of the Project (including, without limitation, any part thereof being or to be executed by the Employer himself or by persons employed or otherwise engaged by the Employer as referred to in clause 29) is materially adversely affected by any act, omission or default of the Contractor, his servants or agents, or any sub-contractor employed by the Contractor, the Employer shall within a reasonable time of such material adverse effect becoming apparent give written notice thereof to the Contractor, and any loss and/or expense thereby caused to the Employer (whether suffered or incurred by the Employer himself or by persons employed or otherwise engaged by the Employer as referred to in clause 29) may be deducted from any monies due or to become due to the Contractor or may be recoverable from the Contractor as a debt. The notice to commence works on site may be issued within [   ] days of the date of formation of the Contract, and shall specify a Date of Access not less than [   ] days and not more than [   ] days after the date of issue of such notice. It shall be within the absolute discretion of the Employer whether or not notice to commence works on site shall be given. Time shall be of the essence with regard to the time limit for the issue of the notice to commence works on site.'

Clause 2.2.1 shall be deleted.

Clause 13: Variations and provisional sums

Delete clause 13.1.3, and substitute:

'.3   compliance with the [Architect's] **OR** [Construction Manager's] instructions under clause 2.1 or clause 29.1 [and nomination of a Sub-Contractor to supply and fix materials or goods or to execute work described in the Contract Documents which has been priced by the Contractor for supply and fixing or execution by the Contractor].'

Clause 18: Partial possession by Employer

In clause 18.1, delete 'and the consent of the Contractor (which consent shall not be unreasonably withheld) has been obtained....'.

Clauses 19.1.1 and 19.1.2: Assignment

Delete clause 19.1.1 and the first sentence of clause 19.1.2, and substitute:

'19.1.1 The Contractor shall not, without the written consent of the Employer, assign this Contract.

19.1.2 Where clause 19.1.2 is stated in the Appendix to apply, the Employer may assign this Contract.'

[In clause 19.2, delete 'other than a Nominated Sub-Contractor', and add:

'All references in the Contract Documents to Nominated Sub-Contractors shall be disregarded'.]

[Delete clause 19.5.1, and substitute:

'Save as otherwise expressed in the Conditions, the Contractor shall remain wholly responsible for carrying out and completing the Works in all respects in accordance with clause 2.1, notwithstanding the sub-contracting of the supply or fixing of any materials or goods, or of the execution of any work'.]

[Clause 19.5.2 shall be deleted.]

Clause 22A: Erection of new buildings – All Risks Insurance of the Works by the Contractor

Clause 22A shall be deleted.

Clause 22B: Erection of new buildings – All Risks Insurance of the Works by the Employer

In clause 22B.1, delete 'Works', and substitute 'Project'.

Clause 22C: Insurance of existing structures – Insurance of Works in or extensions to existing structures

In clause 22C.2, delete 'Works', and substitute 'Project'.

Clause 23: Date of Possession, completion and postponement

Clause 23.1.1 shall be deleted, and the following substituted:

'On the Date of Access, sufficient access to the site shall be given to the Contractor, who shall thereupon begin the Works, regularly and diligently proceed with the same and shall complete the same on or before the Completion Date'.

In clause 23.1.2, delete 'possession', and substitute 'access'.

In clause 23.3.1, delete 'and, subject to clause 18, the Employer shall not be entitled to take possession of any part or parts of the Works until that date'.

Clause 23.3.2 shall be deleted, and the following substituted:

'The Employer may use or occupy the site or the Works or part thereof whether for the purposes of the execution of the balance of the Project (other than the Works), storage of his goods or for any other purpose whatsoever before the date of issue of the certificate of Practical Completion by the [Architect] **OR** [Construction Manager]. Before such use or occupation, the Employer shall notify the insurers under clause 22B or 22C.2 and .4, whichever may be applicable, and obtain confirmation that such use or occupation will not prejudice the insurance'.

Clause 23.3.3 shall be deleted.

Clause 24: Damages for non-completion

[In clause 24.2.1, delete 'as the Employer may require in writing not later than the date of the Final Certificate'.]

OR

[Delete clause 24.2.1, and substitute:

'Subject to the issue of a certificate under clause 24.1, the Contractor shall pay or allow to the

Employer a sum equivalent to any loss or damage suffered or incurred by the Employer and caused by the failure of the Contractor as aforesaid, and the Employer may set-off and deduct the same from any monies due or to become due to the Contractor under this Contract (including any balance stated as due to the Contractor in the Final Certificate) or the Employer may recover the same from the Contractor as a debt'.]

Clause 25.4: Relevant Events

[Clause 25.4.7 shall be deleted.]

Clause 25.4.10 shall be deleted.

Add to clause 25.4 as an additional sub-clause:

'any breach of this Contract by the Employer or any act or omission on the part of the Employer, the [Architect] **OR** [Construction Manager] or the Quantity Surveyor'.

Clause 27: Determination by Employer

Clause 27.2.4 shall be deleted.

The last sentence of clause 27.5.3 shall be deleted.

The proviso to clause 27.6.4.1 shall be deleted.

[Clause 27.7 shall be deleted.]

OR

[In clause 27.7.1, delete '6 months', and substitute '12 (twelve) months'.

In clause 27.7.2, delete '6 month', and substitute '12 (twelve) month'.]

Clause 28A: Determination by Employer or Contractor

Clause 28A.5.5 shall be deleted.

Clause 29: Works by Employer or persons employed or engaged by Employer

Clause 29.1 shall be deleted, and the following substituted:

'In regard to any work forming part of the Project, but not forming part of this Contract, and which is to be carried out by the Employer himself or by persons employed or otherwise engaged by him, the Contractor shall permit, and co-operate with and in, the execution of such work, and shall act in conformity with all reasonable instructions of the [Architect] **OR** [Contract Administrator] relating to such work, its execution and the Contractor's required co-operation'.

In clause 29.2, delete 'Where the Specification does not provide the information referred to in clause 29.1...' and substitute 'Where the Employer requires the execution of work not forming part of the Project and not forming part of this Contract...'.

Clause 30: Certificates and Payments

In clause 30.1.1.2, delete:

'Notwithstanding the fiduciary interest of the Employer in the Retention as stated in clause 30.5.1...'.

Clause 30.5.1 shall be deleted, and the following substituted:

'30.5.1   the Employer's interest in the Retention shall not be fiduciary, either as trustee for the Contractor [or any Nominated Sub-Contractor] or any other person, or in any other capacity; the relationship of the Employer and the Contractor with regard to the Retention shall be solely that of debtor and unsecured creditor, subject to the terms hereof; and the Employer shall have no obligation to invest the Retention or any part thereof;'

Clause 30.5.3 shall be deleted, and the following substituted:

'30.5.3   the Employer shall have no obligation to segregate the Retention or any part thereof in a separate banking account, or in any other manner whatsoever; and shall be entitled to the full beneficial interest in the Retention and every part thereof (and, without limitation, interest thereon and income arising therefrom) unless and until the Retention is paid to the Contractor pursuant to this Contract.'

Clause 30.3 shall be deleted.

In clause 30.9.2, the last word ('either') of the first paragraph shall be deleted.

In clause 30.9.2.1, the last word ('or') shall be deleted.

Clause 30.9.2.2 shall be deleted.

The concluding words of clause 30.9.2 ('whichever shall be the earlier') shall be deleted.

Clause 31: Finance (No. 2) Act 1975 – statutory tax deduction scheme

References to 'the Act', 'the Regulations' or any legislation shall be interpreted to include references to any amendment or re-enactment thereof.

[Clause 35: Nominated Sub-Contractors

Clause 35.18.2 shall be deleted.

In clause 35.21, the word 'not' shall be deleted where it occurs for the second time.

Clause 35.24.9 shall be deleted, and the following substituted:

'35.24.9 The amount properly payable to the Nominated Sub-Contractor under the Sub-Contract resulting from such further nomination shall be included in the amount stated as due in Interim Certificates and added to the Contract Sum. Provided that any extra amount, payable by the Employer in respect of the Sub-Contractor nominated under the further nomination over the price of the Nominated Sub-Contractor who first entered into a sub-contract in respect of the relevant sub-contract works, resulting from such further nomination, may at the time or at any time after such amount is certified in respect of the Sub-Contractor nominated under the further nomination, be deducted by the Employer from monies due or to become due to the Contractor under this Contract, or may be recoverable from the Contractor by the Employer as a debt.']

OR

[Clause 35 shall be deleted.]

Clause 38.4: Provisions relating to clause 38

Clause 38.4.8.1 shall be deleted.

Clause 39.5: Provisions relating to clause 39

Clause 39.5.8.1 shall be deleted.

Clause 40: Use of price adjustment formulae

Clause 40.7.2.1 shall be deleted.

Clause 41: Settlement of Disputes – Arbitration

In clause 41.2.1, add after the opening proviso:

'the Employer and any person or persons employed or otherwise engaged by the Employer as referred to in clause 29, or...'.

Additional Conditions✻

COMMENTARY

Continue as in Form 22, and see the commentary upon that Form.

FORM 25: COLLATERAL WARRANTY BY TRADE CONTRACTOR TO FUNDING INSTITUTION, FOR USE AS AN ANNEXE TO FORMS 22–24

ANNEXE 'A'

DATED _____ **199**

[   ]

('the Trade Contractor')

-and-

[   ]

('the Employer')

-and-

[   ]

('the Fund')

COLLATERAL WARRANTY BY TRADE CONTRACTOR TO FUNDING INSTITUTION

in respect of the construction of

[   ]

✳ An asterisk indicates that there is a relevant note in the commentary upon this Form.

THIS AGREEMENT is made the   day of   One thousand nine hundred and ninety-

BETWEEN:

(1)

[of] **OR** [whose registered office is at]

('the Trade Contractor');

(2)

[of] **OR** [whose registered office is at]

('the Employer'); and

(3)

[whose registered office is at]

('the Fund', which term shall include its successors and assigns).

WHEREAS:

(A)   The Fund has entered into an agreement ('the Finance Agreement') with the Employer for the provision of certain finance in connection with the carrying out of a project of development briefly described as

at

('the Development').

[The Fund entered into the Finance Agreement, and enters into this Agreement, on its own behalf and as agent for a syndicate of banks. Each of the banks which are members of the syndicate from time to time, including banks joining the syndicate after the date of this Agreement, shall be entitled to the benefit of this Agreement in addition to the Fund.]

(B)   The Employer has entered into a contract dated   199   ('the Trade Contract') with the Contractor for the construction of part of the Development.

NOW in consideration of £1 (one pound) paid by the Fund to the Trade Contractor (receipt of which the Trade Contractor hereby acknowledges) **THIS DEED WITNESSETH** as follows:

1.   The Trade Contractor covenants with the Fund that it has duly performed and observed, and will continue duly to perform and observe, all the terms of the Trade Contract on the Trade Contractor's part to be performed and observed and, without prejudice to the generality of the foregoing, the Trade Contractor warrants that it has exercised and will continue to exercise reasonable skill, care and diligence in the performance of its duties to the Employer under the Trade Contract.

2.1   Without prejudice to the generality of clause 1, the Trade Contractor further warrants:

    2.1.1   that it has not used or specified and will not use or specify for use;

    2.1.2   that it has exercised and will continue to exercise reasonable skill, care and diligence to see that there are not used;

    2.1.3   that it is not aware and has no reason to suspect or believe that there have been or will be used; and

    2.1.4   that it will promptly notify the Fund in writing if it becomes aware or has reason to suspect or believe that there have been or will be used;

    in or in connection with the Development, any of the materials or substances identified in clause 2.2.

2.2   The said materials or substances are:

    2.2.1   high alumina cement in structural elements;

    2.2.2   wood wool slabs in permanent formwork to concrete;

    2.2.3   calcium chloride in admixtures for use in reinforced concrete;

    2.2.4   asbestos products;

    2.2.5   naturally occurring aggregates for use in reinforced concrete which do not comply with British Standard 882: 1983 and/or naturally occurring aggregates for use in concrete which do not comply with British Standard 8110: 1985.

3.   The Fund has no authority to issue any direction or instruction to the Trade Contractor

in relation to performance of the Trade Contractor's duties under the Trade Contract unless and until the Fund has given notice under clauses 5 or 6.

4.   The Trade Contractor acknowledges that the Employer has paid all sums due and owing to the Trade Contractor under the Trade Contract up to the date of this Agreement. The Fund has no liability to the Trade Contractor in respect of sums due under the Trade Contract unless and until the Fund has given notice under clauses 5 or 6.

5.   The Trade Contractor agrees that, in the event of the termination of the Finance Agreement by the Fund, or the occurrence of any event of default under the Finance Agreement, the Trade Contractor will, if so required by notice in writing given by the Fund and subject to clause 7, accept the instructions of the Fund or its appointee to the exclusion of the Employer in respect of the Development upon the terms and conditions of the Trade Contract. The Employer acknowledges that the Trade Contractor shall be entitled to rely on a notice given to the Trade Contractor by the Fund under this clause 5 as conclusive evidence for the purposes of this Agreement of the termination of the Finance Agreement by the Fund, or of the occurrence of any such event of default.

6.   The Trade Contractor further agrees that it will not, without first giving the Fund not less than 21 (twenty-one) days' notice in writing, exercise any right it may have to terminate the Trade Contract or determine its employment thereunder, or to treat the same as having been repudiated by the Employer, or discontinue the performance of any duties to be performed by the Trade Contractor pursuant thereto for any reason whatsoever (including, without limitation, any act or omission by or on behalf of the Employer). The Trade Contractor's right to terminate the Trade Contract or determine its employment thereunder, or treat the same as having been repudiated, or discontinue performance shall cease if, within such period of notice and subject to clause 7, the Fund shall give notice in writing to the Trade Contractor requiring the Trade Contractor to accept the instructions of the Fund or its appointee to the exclusion of the Employer in respect of the Development upon the terms and conditions of the Trade Contract.

7.   It shall be a condition of any notice given by the Fund under clauses 5 or 6 that the Fund or its appointee accepts liability for payment of all sums payable to the Trade Contractor under the Trade Contract and for performance of the Employer's obligations under the Trade Contract, including payment of any sums outstanding at the date of such notice, but excluding any such sums in respect of which funds were advanced to the Employer pursuant to the Finance Agreement before the date of such notice. Upon the issue of any notice by the Fund under clauses 5 or 6, the Trade Contract shall continue in full force and effect as if no right of termination or

determination on the part of the Trade Contractor had arisen and the Trade Contractor shall be liable to the Fund or its appointee under the Trade Contract in lieu of its liability to the Employer. If any notice given by the Fund under clauses 5 or 6 requires the Trade Contractor to accept the instructions of the Fund's appointee, the Fund shall be liable to the Trade Contractor as guarantor for the payment of all sums from time to time due to the Trade Contractor from the Fund's appointee.

8.   The copyright in all drawings, reports, specifications, bills of quantities, calculations and other similar documents provided by the Trade Contractor in connection with the Development shall remain vested in the Trade Contractor, but the Fund and its appointee shall have a licence to copy and use such drawings and other documents, and to reproduce the designs contained in them, for any purpose related to the Development including, but without limitation, the construction, completion, maintenance, letting, promotion, advertisement, reinstatement, repair and/or extension of the Development. The Trade Contractor shall, if the Fund so requests and undertakes in writing to pay the Trade Contractor's reasonable copying charges, promptly supply the Fund with conveniently reproducible copies of all such drawings and other documents.

9.1   The Trade Contractor shall maintain professional indemnity insurance covering (inter alia) all liability hereunder upon customary and usual terms and conditions prevailing for the time being in the insurance market, and with reputable insurers lawfully carrying on such insurance business in the United Kingdom, in an amount of not less than   pounds (£   ) for any one occurrence or series of occurrences arising out of any one event for a period beginning now and ending 15 (fifteen) years after the date of practical completion of the Development, provided always that such insurance is available at commercially reasonable rates. The said terms and conditions shall not include any term or condition to the effect that the Trade Contractor must discharge any liability before being entitled to recover from the insurers, or any other term or condition which might adversely affect the rights of any person to recover from the insurers pursuant to the Third Parties (Rights Against Insurers) Act 1930, or any amendment or re-enactment thereof. The Trade Contractor shall not, without the prior approval in writing of the Fund, settle or compromise with the insurers any claim which the Trade Contractor may have against the insurers and which relates to a claim by the Fund against the Trade Contractor, or by any act or omission lose or prejudice the Trade Contractor's right to make or proceed with such a claim against the insurers.

9.2   Any increased or additional premium required by insurers by reason of the Trade Contractor's own claims record or other acts, omissions, matters or things particular to the Trade Contractor shall be deemed to be within commercially reasonable rates.

9.3   The Trade Contractor shall immediately inform the Fund if such insurance ceases to be available at commercially reasonable rates in order that the Trade Contractor and the Fund can discuss means of best protecting the respective positions of the Fund and the Trade Contractor in respect of the Development in the absence of such insurance.

9.4   The Trade Contractor shall fully co-operate with any measures reasonably required by the Fund, including (without limitation) completing any proposals for insurance and associated documents, maintaining such insurance at rates above commercially reasonable rates if the Fund undertakes in writing to reimburse the Trade Contractor in respect of the net cost of such insurance to the Trade Contractor above commercially reasonable rates or, if the Fund effects such insurance at rates at or above commercially reasonable rates, reimbursing the Fund in respect of what the net cost of such insurance to the Fund would have been at commercially reasonable rates.

9.5   As and when it is reasonably requested to do so by the Fund or its appointee under clauses 5 or 6, the Trade Contractor shall produce for inspection documentary evidence (including, if required by the Fund or such appointee, the original of the relevant insurance documents) that its professional indemnity insurance is being maintained.

10.   The Employer has agreed to be a party to this Agreement for the purpose of clause 12 and for acknowledging that the Trade Contractor shall not be in breach of the Trade Contract by complying with the obligations imposed on it by this Agreement.

11.   This Agreement may be assigned by the Fund and its successors and assigns without the consent of the Employer or the Trade Contractor being required.

12.   The Employer and the Trade Contractor undertake with the Fund not to vary, or depart from, the terms and conditions of the Trade Contract without the prior written consent of the Fund, and agree that no such variation or departure made without such consent shall be binding on the Fund, or affect or prejudice the Fund's rights hereunder, or under the Trade Contract, or in any other way.

13.   Any notice to be given by the Trade Contractor hereunder shall be deemed to be duly given if it is delivered by hand at or sent by registered post or recorded delivery to the Fund at its registered office, and any notice to be given by the Fund hereunder shall be deemed to be duly given if it is addressed to the Trade Contractor and delivered by hand at or sent by registered post or recorded delivery to the above-mentioned address of the Trade Contractor or to the principal business address of the Trade Contractor for the time being and, in the case of any such notices, the same shall if sent by registered post or recorded delivery be deemed to have been received forty-eight hours after being posted.

IN WITNESS whereof the Trade Contractor and the Employer have executed this Deed on the date first stated above.✳

COMMENTARY

Cf. Forms 6 and 19.

FORM 26: COLLATERAL WARRANTY BY TRADE CONTRACTOR TO ACQUIRER, FOR USE AS AN ANNEXE TO FORMS 22–24

ANNEXE 'B'

DATED _____ **199**

[   ]

('the Trade Contractor')

-and-

[   ]

('the Employer')

-and-

[   ]

('the Acquirer')

COLLATERAL WARRANTY BY TRADE CONTRACTOR TO ACQUIRER

in respect of the construction of

[   ]

✱ An asterisk indicates that there is a relevant note in the commentary upon this Form.

THIS AGREEMENT is made the   day of   One thousand nine hundred and ninety-

BETWEEN:

(1)

[of] **OR** [whose registered office is at]

('the Trade Contractor');

(2)

[of] **OR** [whose registered office is at]

('the Employer'); and

(3)

[of] **OR** [whose registered office is at]

('the Acquirer', which term shall include its successors and assigns).

WHEREAS:

(A)   The Acquirer intends to acquire, or has acquired, an interest in a project of development briefly described as

at

('the Development').

(B)   The Employer has entered into a contract dated   199   ('the Trade Contract') with the Trade Contractor for the construction of part of the Development.

NOW in consideration of £1 (one pound) paid by the Acquirer to the Trade Contractor (receipt of which the Trade Contractor hereby acknowledges) **THIS DEED WITNESSETH** as follows:

1.   The Trade Contractor covenants with the Acquirer that it has duly performed and observed, and will continue duly to perform and observe, all the terms of the Trade Contract on the Trade Contractor's part to be performed and observed and, without

prejudice to the generality of the foregoing, the Trade Contractor warrants that it has exercised and will continue to exercise reasonable skill, care and diligence in the performance of its duties to the Employer under the Trade Contract.

2.1   Without prejudice to the generality of clause 1, the Trade Contractor further warrants:

        2.1.1   that it has not used or specified and will not use or specify for use;

        2.1.2   that it has exercised and will continue to exercise reasonable skill, care and diligence to see that there are not used;

        2.1.3   that it is not aware and has no reason to suspect or believe that there have been or will be used; and

        2.1.4   that it will promptly notify the Acquirer in writing if it becomes aware or has reason to suspect or believe that there have been or will be used;

in or in connection with the Development, any of the materials or substances identified in clause 2.2.

2.2   The said materials or substances are:

        2.2.1   high alumina cement in structural elements;

        2.2.2   wood wool slabs in permanent formwork to concrete;

        2.2.3   calcium chloride in admixtures for use in reinforced concrete;

        2.2.4   asbestos products;

        2.2.5   naturally occurring aggregates for use in reinforced concrete which do not comply with British Standard 882: 1983 and/or naturally occurring aggregates for use in concrete which do not comply with British Standard 8110: 1985.

3.   The Acquirer has no authority to issue any direction or instruction to the Trade Contractor in relation to performance of the Trade Contractor's duties under the Trade Contract.

4.   The Trade Contractor acknowledges that the Employer has paid all sums due and owing to the Trade Contractor under the Trade Contract up to the date of this Agreement. The Acquirer has no liability to the Trade Contractor in respect of sums due under the Trade Contract.

5.   The copyright in all drawings, reports, specifications, bills of quantities, calculations and other similar documents provided by the Trade Contractor in connection with the Development shall remain vested in the Trade Contractor, but the Acquirer and its appointee shall have a licence to copy and use such drawings and other documents, and to reproduce the designs contained in them, for any purpose related to the Development including, but without limitation, the construction, completion, maintenance, letting, promotion, advertisement, reinstatement, repair and/or extension of the Development. The Trade Contractor shall, if the Acquirer so requests and undertakes in writing to pay the Trade Contractor's reasonable copying charges, promptly supply the Acquirer with conveniently reproducible copies of all such drawings and other documents.

6.1   The Trade Contractor shall maintain professional indemnity insurance covering (inter alia) all liability hereunder upon customary and usual terms and conditions prevailing for the time being in the insurance market, and with reputable insurers lawfully carrying on such insurance business in the United Kingdom, in an amount of not less than   pounds (£   ) for any one occurrence or series of occurrences arising out of any one event for a period beginning now and ending 15 (fifteen) years after the date of practical completion of the Development, provided always that such insurance is available at commercially reasonable rates. The said terms and conditions shall not include any term or condition to the effect that the Trade Contractor must discharge any liability before being entitled to recover from the insurers, or any other term or condition which might adversely affect the rights of any person to recover from the insurers pursuant to the Third Parties (Rights Against Insurers) Act 1930, or any amendment or re-enactment thereof. The Trade Contractor shall not, without the prior approval in writing of the Acquirer, settle or compromise with the insurers any claim which the Trade Contractor may have against the insurers and which relates to a claim by the Acquirer against the Trade Contractor, or by any act or omission lose or prejudice the Trade Contractor's right to make or proceed with such a claim against the insurers.

6.2   Any increased or additional premium required by insurers by reason of the Trade Contractor's own claims record or other acts, omissions, matters or things particular to the Trade Contractor shall be deemed to be within commercially reasonable rates.

6.3   The Trade Contractor shall immediately inform the Acquirer if such insurance ceases to be available at commercially reasonable rates in order that the Trade Contractor and the Acquirer can discuss means of best protecting the respective positions of the Acquirer and the Trade Contractor in respect of the Development in the absence of such insurance.

6.4   The Trade Contractor shall fully co-operate with any measures reasonably required by the Acquirer, including (without limitation) completing any proposals for insurance and associated documents, maintaining such insurance at rates above commercially reasonable rates if the Acquirer undertakes in writing to reimburse the Trade Contractor in respect of the net cost of such insurance to the Trade Contractor above commercially reasonable rates or, if the Acquirer effects such insurance at rates at or above commercially reasonable rates, reimbursing the Acquirer in respect of what the net cost of such insurance to the Acquirer would have been at commercially reasonable rates.

6.5   As and when it is reasonably requested to do so by the Acquirer the Trade Contractor shall produce for inspection documentary evidence (including, if required by the Acquirer, the original of the relevant insurance documents) that its professional indemnity insurance is being maintained.

7.   The Employer has agreed to be a party to this Agreement for the purpose of clause 9 and for acknowledging that the Trade Contractor shall not be in breach of the Trade Contract by complying with the obligations imposed on it by this Agreement.

8.   This Agreement may be assigned by the Acquirer and its successors and assigns without the consent of the Employer or the Trade Contractor being required.

9.   The Employer and the Trade Contractor undertake with the Acquirer not to vary, or depart from, the terms and conditions of the Trade Contract without the prior written consent of the Acquirer, and agree that no such variation or departure made without such consent shall be binding on the Acquirer, or affect or prejudice the Acquirer's rights hereunder, or in any other way.

10.   Any notice to be given by the Trade Contractor hereunder shall be deemed to be duly given if it is delivered by hand at or sent by registered post or recorded delivery to the above-mentioned address of the Acquirer or to the principal business address of the Acquirer for the time being, and any notice to be given by the Acquirer hereunder shall be deemed to be duly given if it is addressed to the Trade Contractor and delivered by hand at or sent by registered post or recorded delivery to the above-mentioned address of the Trade Contractor or to the principal business address of the Trade Contractor for the time being and, in the case of any such notices, the same shall if sent by registered post or recorded delivery be deemed to have been received forty-eight hours after being posted.

IN WITNESS whereof the Trade Contractor and the Employer have executed this Deed on the date first stated above.✱

COMMENTARY

Cf. Forms 7 and 20.

FORM 27: COLLATERAL WARRANTY BY TRADE CONTRACTOR'S SUB-CONTRACTOR, FOR USE AS AN ANNEXE TO FORMS 22–24

ANNEXE 'D'

DATED   **199**

[   ]

('the Sub-Contractor')

-and-

[   ]

(['the Employer'] **OR** ['the Fund'] **OR** ['the Acquirer'])

COLLATERAL WARRANTY BY
TRADE CONTRACTOR'S SUB-CONTRACTOR

in respect of the construction of
[   ]

✱ **An asterisk indicates that there is a relevant note in the commentary upon this Form.**

THIS AGREEMENT is made the   day of   One thousand nine hundred and ninety-

BETWEEN:

(1)

[of] **OR** [whose registered office is at]

('the Sub-Contractor'); and

(2)

[of] **OR** [whose registered office is at]

(['the Employer'] **OR** ['the Fund'] **OR** ['the Acquirer'], which term shall include its successors and assigns).

WHEREAS:

(A)   [The Employer] **OR** [('the Employer')] has entered into a contract dated   199   ('the Trade Contract') with [   ] ('the Trade Contractor') for the construction of part of a project of development briefly described as

at

('the Development').

(B)   The Trade Contractor has entered into a sub-contract dated   199 ('the Sub-Contract') with the Sub-Contractor for the execution of certain sub-contract works briefly described as

and forming part of the Development ('the Sub-Contract Works').

[(C)   The Fund has entered into an agreement with the Employer for the provision of certain finance in connection with the carrying out of the Development.] [The Fund entered into such agreement, and enters into this Agreement, on its own behalf and as agent for a syndicate of banks. Each of the banks which are members of the syndicate from time to time, including banks joining the syndicate after the date of this Agreement, shall be entitled to the benefit of this Agreement in addition to the Fund.]

OR

[(C)   The Acquirer intends to acquire, or has acquired, an interest in the Development.]

NOW in consideration of £1 (one pound) paid by the [Employer] **OR** [Fund] **OR** [Acquirer] to the Sub-Contractor (receipt of which the Sub-Contractor hereby acknowledges) **THIS DEED WITNESSETH** as follows:

1.   The Sub-Contractor covenants with the [Employer] **OR** [Fund] **OR** [Acquirer] that it has duly performed and observed, and will continue duly to perform and observe, all the terms of the Sub-Contract on the Sub-Contractor's part to be performed and observed and, without prejudice to the generality of the foregoing, the Sub-Contractor warrants that it has exercised and will continue to exercise reasonable skill, care and diligence in the performance of its duties to the Trade Contractor under the Sub-Contract.

2.1   Without prejudice to the generality of clause 1, the Sub-Contractor further warrants:

   2.1.1   that it has not used or specified and will not use or specify for use;

   2.1.2   that it has exercised and will continue to exercise reasonable skill, care and diligence to see that there are not used;

   2.1.3   that it is not aware and has no reason to suspect or believe that there have been or will be used; and

   2.1.4   that it will promptly notify the [Employer] **OR** [Fund] **OR** [Acquirer] in writing if it becomes aware or has reason to suspect or believe that there have been or will be used;

in or in connection with the Development, any of the materials or substances identified in clause 2.2.

2.2   The said materials or substances are:

   2.2.1   high alumina cement in structural elements;

   2.2.2   wood wool slabs in permanent formwork to concrete;

   2.2.3   calcium chloride in admixtures for use in reinforced concrete;

   2.2.4   asbestos products;

2.2.5   naturally occurring aggregates for use in reinforced concrete which do not comply with British Standard 882: 1983 and/or naturally occurring aggregates for use in concrete which do not comply with British Standard 8110: 1985.

3.   The [Employer] **OR** [Fund] **OR** [Acquirer] has no authority to issue any direction or instruction to the Sub-Contractor in relation to performance of the Sub-Contractor's duties under the Sub-Contract.

4.   The Sub-Contractor acknowledges that the Trade Contractor has paid all sums due and owing to the Sub-Contractor under the Sub-Contract up to the date of this Agreement. The [Employer] **OR** [Fund] **OR** [Acquirer] has no liability to the Sub-Contractor in respect of sums due under the Sub-Contract.

5.   The copyright in all drawings, reports, specifications, bills of quantities, calculations and other similar documents provided by the Sub-Contractor in connection with the Development shall remain vested in the Sub-Contractor (or as may be otherwise provided by the Sub-Contract), but the [Employer] **OR** [Fund] **OR** [Acquirer] and its appointee shall have a licence to copy and use such drawings and other documents, and to reproduce the designs contained in them, for any purpose related to the Development including, but without limitation, the construction, completion, maintenance, letting, promotion, advertisement, reinstatement, repair and/or extension of the Development. The Sub-Contractor shall, if the [Employer] **OR** [Fund] **OR** [Acquirer] so requests and undertakes in writing to pay the Sub-Contractor's reasonable copying charges, promptly supply the [Employer] **OR** [Fund] **OR** [Acquirer] with conveniently reproducible copies of all such drawings and other documents.

6.1   The Sub-Contractor shall maintain professional indemnity insurance covering (inter alia) all liability hereunder upon customary and usual terms and conditions prevailing for the time being in the insurance market, and with reputable insurers lawfully carrying on such insurance business in the United Kingdom, in an amount of not less than   pounds (£   ) for any one occurrence or series of occurrences arising out of any one event for a period beginning now and ending 15 (fifteen) years after the date of practical completion of the Development, provided always that such insurance is available at commercially reasonable rates. The said terms and conditions shall not include any term or condition to the effect that the Sub-Contractor must discharge any liability before being entitled to recover from the insurers, or any other term or condition which might adversely affect the rights of any person to recover from the insurers pursuant to the Third Parties (Rights Against Insurers) Act 1930, or any amendment or re-enactment thereof. The Sub-Contractor shall not, without the prior approval in writing of the [Employer] **OR** [Fund] **OR** [Acquirer], settle or compromise with the insurers any claim which the Sub-

Contractor may have against the insurers and which relates to a claim by the [Employer] **OR** [Fund] **OR** [Acquirer] against the Sub-Contractor, or by any act or omission lose or prejudice the Sub-Contractor's right to make or proceed with such a claim against the insurers.

6.2   Any increased or additional premium required by insurers by reason of the Sub-Contractor's own claims record or other acts, omissions, matters or things particular to the Sub-Contractor shall be deemed to be within commercially reasonable rates.

6.3   The Sub-Contractor shall immediately inform the [Employer] **OR** [Fund] **OR** [Acquirer] if such insurance ceases to be available at commercially reasonable rates in order that the Sub-Contractor and the [Employer] **OR** [Fund] **OR** [Acquirer] can discuss means of best protecting the respective positions of the [Employer] **OR** [Fund] **OR** [Acquirer] and the Sub-Contractor in respect of the Development in the absence of such insurance.

6.4   The Sub-Contractor shall fully co-operate with any measures reasonably required by the [Employer] **OR** [Fund] **OR** [Acquirer], including (without limitation) completing any proposals for insurance and associated documents, maintaining such insurance at rates above commercially reasonable rates if the [Employer] **OR** [Fund] **OR** [Acquirer] undertakes in writing to reimburse the Sub-Contractor in respect of the net cost of such insurance to the Sub-Contractor above commercially reasonable rates or, if the [Employer] **OR** [Fund] **OR** [Acquirer] effects such insurance at rates at or above commercially reasonable rates, reimbursing the [Employer] **OR** [Fund] **OR** [Acquirer] in respect of what the net cost of such insurance to the [Employer] **OR** [Fund] **OR** [Acquirer] would have been at commercially reasonable rates.

6.5   As and when it is reasonably requested to do so by the [Employer] **OR** [Fund] **OR** [Acquirer] the Sub-Contractor shall produce for inspection documentary evidence (including, if required by the [Employer] **OR** [Fund] **OR** [Acquirer], the original of the relevant insurance documents) that its professional indemnity insurance is being maintained.

7.   This Agreement may be assigned by the [Employer] **OR** [Fund] **OR** [Acquirer] and its successors and assigns without the consent of the Sub-Contractor being required.

8.   Any notice to be given by the Sub-Contractor hereunder shall be deemed to be duly given if it is delivered by hand at or sent by registered post or recorded delivery to the above-mentioned address of the [Employer] **OR** [Fund] **OR** [Acquirer] or to the principal business address of the [Employer] **OR** [Fund] **OR** [Acquirer] for the time being, and any notice to be given by the [Employer] **OR** [Fund] **OR** [Acquirer] hereunder shall be deemed to be duly given if it is addressed to the Sub-Contractor and

delivered by hand at or sent by registered post or recorded delivery to the above-mentioned address of the Sub-Contractor or to the principal business address of the Sub-Contractor for the time being and, in the case of any such notices, the same shall if sent by registered post or recorded delivery be deemed to have been received forty-eight hours after being posted.

NOTE: The following clause will only be required in the case of Collateral Warranties in favour of the Employer or Fund.

9.   The Sub-Contractor shall within 7 (seven) working days of the [Employer's] **OR** [Fund's] request to so so, execute, in favour of any persons who have entered or shall enter into an agreement for the provision of finance in connection with the Development and/or in favour of any persons who have acquired or shall acquire any interest in or over the Development or any part thereof, a Deed in the form of this Deed, excluding this clause, or a similar form reasonably required by the [Employer] **OR** [Fund], and deliver the same duly executed to the [Employer] **OR** [Fund]; together in each case (if requested by the [Employer] **OR** [Fund]) with a guarantee (in form and substance reasonably required by the [Employer] **OR** [Fund]) from the ultimate parent company of the Sub-Contractor in respect of the Sub-Contractor's obligations pursuant to such Deed.

IN WITNESS whereof the Sub-Contractor has executed this Deed on the date first stated above.✱

COMMENTARY

Cf. Form 21.

FORM 28: PARENT COMPANY GUARANTEE FOR THE PERFORMANCE OF CONTRACTOR'S COLLATERAL WARRANTY, FOR USE AS AN ANNEXE TO FORMS 12–18 AND 22–24

ANNEXE 'C'

DATED _____ **199**

[   ]

('the Guarantor')

-and-

[   ]

[('the Fund) **OR** ('the Acquirer')]

GUARANTEE

in respect of a Contractor's Collateral Warranty
relating to the construction of
[   ]

✻ An asterisk indicates that there is a relevant note in the commentary upon this Form.

THIS AGREEMENT is made the   day of   One thousand nine hundred and ninety-

BETWEEN:

(1)

[of] **OR** [whose registered office is at]

('the Guarantor'); and

(2)

[of] **OR** [whose registered office is at]

(['the Fund'] **OR** ['the Acquirer'], which term shall include its successors and assigns).

WHEREAS by an Agreement ('the Collateral Warranty') dated   199 and made between [   ] ('the Contractor') [(as Employer)] and [the Fund] **OR** [the Acquirer], the Contractor assumed certain obligations towards the [Fund] **OR** [Acquirer].

NOW THIS DEED WITNESSETH that if the Contractor defaults in the discharge of any of the Contractor's obligations under or pursuant to the Collateral Warranty, the Guarantor will indemnify the [Fund] **OR** [Acquirer] against all loss and damage thereby caused to the [Fund] **OR** [Acquirer], and no alterations in the Collateral Warranty and no extension of time, forbearance or forgiveness, nor any act, matter or thing whatsoever except an express release by Deed by the [Fund] **OR** [Acquirer], shall in any way release the Guarantor from any liability hereunder.

IN WITNESS whereof the Guarantor has executed this Deed on the date first stated above.✱

COMMENTARY

Cf. Form 9.

FORM 29: PARENT COMPANY GUARANTEE FOR THE PERFORMANCE OF SUB-CONTRACTOR'S COLLATERAL WARRANTY, FOR USE AS AN ANNEXE TO FORMS 12–17 AND 22–24

ANNEXE ' '

DATED _____ **199**

[   ]

('the Guarantor')

-and-

[   ]

(['the Employer'] **OR** ['the Fund'] **OR** ['the Acquirer'])

GUARANTEE

in respect of a Sub-Contractor's
Collateral Warranty relating to the construction of
[   ]

✱ An asterisk indicates that there is a relevant note in the commentary upon this Form.

THIS AGREEMENT is made the   day of   One thousand nine hundred and ninety-

BETWEEN:

(1)

[of] **OR** [whose registered office is at]

('the Guarantor'); and

(2)

[of] **OR** [whose registered office is at]

(['the Employer'] **OR** ['the Fund'] **OR** ['the Acquirer']), which term shall include its successors and assigns).

WHEREAS by an Agreement ('the Collateral Warranty') dated   199 and made between [   ] ('the Sub-Contractor') and [the Employer] **OR** [the Fund] **OR** [the Acquirer], the Sub-Contractor assumed certain obligations towards the [Employer] **OR** [Fund] **OR** [Acquirer].

NOW THIS DEED WITNESSETH that if the Sub-Contractor defaults in the discharge of any of the Sub-Contractor's obligations under or pursuant to the Collateral Warranty, the Guarantor will indemnify the [Employer] **OR** [Fund] **OR** [Acquirer] against all loss and damage thereby caused to the [Employer] **OR** [Fund] **OR** [Acquirer], and no alterations in the Collateral Warranty, and no extension of time, forbearance or forgiveness, nor any act, matter or thing whatsoever except an express release by Deed by the [Employer] **OR** [Fund] **OR** [Acquirer], shall in any way release the Guarantor from any liability hereunder.

IN WITNESS whereof the Guarantor has executed this Deed on the date first stated above.✶

COMMENTARY

Cf. Form 9

318

FORM 30: PERFORMANCE BOND, FOR USE AS AN ANNEXE TO FORMS 12–14 AND 22–24

ANNEXE 'F'

DATED _____ **199**

[   ]

('the Contractor')

-and-

[   ]

('the Surety')

-and-

[   ]

('the Employer')

PERFORMANCE BOND

in respect of the construction of
[   ]

✱An asterisk indicates that there is a relevant note in the commentary upon this Form.

THIS BOND is made the   day of   One thousand nine hundred and ninety-

BETWEEN:

(1)

> [of] **OR** [whose registered office is at]

> ('the Contractor');

(2)

> [of] **OR** [whose registered office is at]

> ('the Surety'); and

(3)

> [of] **OR** [whose registered office is at]

> ('the Employer', which term shall include its successors and assigns).

WHEREAS by an Agreement ('the Contract') dated   199   and made between the Employer of the one part and the Contractor of the other part, the Contractor undertook the construction of certain Works in accordance with the terms and conditions of the Contract.

NOW THIS DEED WITNESSETH as follows:

1   Bond

By this Bond the Contractor and the Surety, their successors and assigns, are jointly and severally held and bound to the Employer for payment to the Employer of the sum of   pounds (£   ).

2   Conditions

The conditions of this Bond are that if:

2.1   the Contractor duly discharges all the Contractor's obligations under or pursuant to the Contract; or

2.2   in the event of the Contractor's default in the discharge of any such obligations, the Surety shall pay to the Employer the loss and damage thereby caused to the Employer, up to the amount of this Bond; or

2.3   pursuant to clause 30.8 of the Contract, any balance due from the Contractor to the Employer pursuant to a Final Certificate (having been adjusted, if necessary, by agreement between the parties or as a result of arbitration or litigation) has been paid, and (without any undetermined or unresolved exception as provided in clauses 30.9.2 or 30.9.3 of the Contract, and save in respect of fraud) such Final Certificate would, pursuant to clause 30.9.1 of the Contract, have effect in any proceedings arising out of, or in connection with, the Contract (whether by arbitration under Article 5 thereof or otherwise) as conclusive evidence of the matters described in clauses 30.9.1.1, 30.9.1.2, 30.9.1.3 and 30.9.1.4 of the Contract; and any amount due from the Contractor to the Employer pursuant to any award or judgment in, or settlement of, any arbitration or other proceedings commenced in respect of the Contract before or within 28 (twenty-eight) days after the said Final Certificate was issued, has been paid;

this Bond shall thereby be discharged, but otherwise shall remain in force.

3   Alterations

No alterations in the Contract, or in the Works, and no extension of time, forbearance or forgiveness, nor any act, matter or thing whatsoever except fulfilment of one of the above conditions or an express release by Deed by the Employer, shall in any way release the Surety from any liability under this Bond.

[4   Reduction on Practical Completion]

[Provided that upon certification of Practical Completion of the Works under the Contract, the amount of this Bond shall reduce by one-half]**OR**[Provided that upon certification of Practical Completion of each Section of the Works under the Contract, the amount of this Bond shall reduce by the following respective amounts:

£

Section 1

Section 2   ]

IN WITNESS whereof the Contractor and the Surety have executed this Deed on the date first stated above.✲

COMMENTARY

See paragraphs 38 and 39, Chapter 1.

FORM 31: PARENT COMPANY GUARANTEE FOR THE PERFORMANCE OF A CONTRACT, FOR USE AS AN ANNEXE TO FORMS 12–14, 17, 18 AND 22–24

ANNEXE ' '

DATED _____ **199**

[   ]

('the Guarantor')

-and-

[   ]

('the Employer')

GUARANTEE

in respect of the construction of

[   ]

✱ An asterisk indicates that there is a relevant note in the commentary upon this Form.

THIS AGREEMENT is made the   day of   One thousand nine hundred and ninety-

BETWEEN:

(1)

[of] **OR** [whose registered office is at]

('the Guarantor'); and

(2)

[of] **OR** [whose registered office is at]

('the Employer', which term shall include its successors and assigns).

WHEREAS by an Agreement ('the Contract') dated   199   and made between the Employer of the one part and [   ] ('the Contractor') of the other part, the Contractor undertook the construction of certain Works in accordance with the terms and conditions of the Contract.

NOW THIS DEED WITNESSETH that if the Contractor defaults in the discharge of any of the Contractor's obligations under or pursuant to the Contract, the Guarantor will indemnify the Employer against all loss and damage thereby caused to the Employer, and no alterations in the Contract, or in the Works, and no extension of time, forbearance or forgiveness, nor any act, matter or thing whatsoever except an express release by Deed by the Employer, shall in any way release the Guarantor from any liability hereunder.

IN WITNESS whereof the Guarantor has executed this Deed on the date first stated above.✳

COMMENTARY

See paragraphs 39 and 40, Chapter 1.

FORM 32: NOVATION AGREEMENT

[   ]

('the Employer')

-and-

[   ]

('the Original Contractor')

-and-

[   ]

('the Substituted Contractor')

NOVATION AGREEMENT

in respect of the construction of
[   ]

✱ An asterisk indicates that there is a relevant note in the commentary upon this Form.

THIS AGREEMENT is made the   day of   One thousand nine hundred and ninety-

BETWEEN:

(1)

[of] **OR** [whose registered office is at]

('the Employer', which term shall include its successors and assigns);

(2)

[of] **OR** [whose registered office is at]

('the Original Contractor'); and

(3)

[of] **OR** [whose registered office is at]

('the Substituted Contractor').

WHEREAS:

(A)   This Agreement is supplemental to a Main Contract, short particulars of which are set out in the Schedule hereto.

(B)   The Original Contractor has requested the Employer to accept the Substituted Contractor in the place of the Original Contractor in respect of the Main Contract, which the Employer has agreed to do upon the terms hereof.

NOW THIS DEED WITNESSETH as follows:

1.   The Substituted Contractor covenants with the Employer, and the Employer covenants with the Substituted Contractor, to perform and observe the Main Contract and to be bound thereby as if the Substituted Contractor had been a party to the Main Contract 'ab initio' in the place of the Original Contractor.

2.   The Original Contractor hereby warrants to the Substituted Contractor that the Original Contractor has duly performed and observed all its obligations pursuant to the Main Contract prior to the date hereof.

3.   The Employer releases and discharges the Original Contractor, and the Original Contractor releases and discharges the Employer, from the Main Contract and all actions, proceedings, costs, claims, demands and liabilities in respect thereof. The Employer agrees with the Original Contractor and the Substituted Contractor to accept the performance and observance of the Main Contract by the Substituted Contractor in lieu of such performance and observance by the Original Contractor.

IN WITNESS whereof the parties have executed this Deed in triplicate on the date first stated above.✱

SCHEDULE

Particulars of Main Contract

Date		**Parties**	**Description of Works and Site**
199	(1)	The Employer	
	(2)	The Original Contractor	

COMMENTARY

See paragraph 41, Chapter 1. The effect of novation on the related obligations of third parties – for example, performance bonds or parent company guarantees in respect of the Contractor's performance of the Main Contract – should be considered, and fresh obligations obtained from such third parties, if appropriate.

APPENDICES

APPENDICES

APPENDIX A

DOCUMENTARY FORMALITIES AND ATTESTATION OF DEEDS

Execution of Deeds An appropriate attestation clause for execution of a Deed by an individual is as follows:

SIGNED as a **DEED** by   )
JOHN SMITH in the   )   **JOHN SMITH**
presence of:)

ANDREW PIKE
14 Dominion Street,
London EC2M 2RJ

Solicitor

      A witness to a Deed should add his or her name, address and occupation or description, as illustrated.
      If a partnership is executing a Deed, all the partners should sign as above, or an authorized partner or authorized partners may sign on behalf of the partnership as follows:

SIGNED as a **DEED** by   )
JOHN SMITH for and on)
behalf of **THE JOHN**   )   **JOHN SMITH**
SMITH PARTNERSHIP in   )
the presence of:)

ANDREW PIKE
14 Dominion Street
London EC2M 2RJ

Solicitor

      Unless all partners sign, those signing should prove their authority to execute Deeds on behalf of the partnership, e.g. by a power of attorney or certified extract from their partnership agreement.

A company executing a Deed using a common seal could execute as follows:

THE COMMON SEAL of)
JOHN SMITH [LIMITED] OR)
[PLC] was affixed to)   **[COMMON SEAL]**
this **DEED** in the)
presence of:)

      JOHN SMITH   Director

      JANE JONES   Secretary

A company not using a common seal could execute as follows:

SIGNED as a **DEED** for)
and on behalf of **JOHN**)
SMITH [LIMITED] OR)
[PLC] by:)

      JOHN SMITH   Director

      JANE JONES   Secretary

The above attestation clauses conform with the law on the execution of Deeds as amended with effect from 31 July 1990 by section 1 of the Law of Property (Miscellaneous Provisions) Act 1989 and section 130(2) of the Companies Act 1989. The principal changes from the previous practice are that:

— if a document is to be a Deed it must be clear **on its face** that it is intended to be a Deed;
— sealing by individuals (including partners) has been abolished;
— a company may choose whether or not to use a common seal.

Attestation clauses should preferably be added at the very end of the relevant document, i.e. after any Schedules, Appendices, etc.

If it becomes necessary to execute a document otherwise than as a Deed, all references to 'Deed' in the relevant text should be replaced by 'Agreement', and the attestation clauses should be as follows:

SIGNED by   )
JOHN SMITH in the   )
presence of:)

OR

SIGNED by **JOHN SMITH**   )
for and on behalf of **THE**   )
JOHN SMITH PARTNERSHIP)
in the presence of:)

OR

SIGNED by **JOHN SMITH**   )
for and on behalf of   )
JOHN SMITH [LIMITED]   )
OR [PLC] in the presence)
of:)

APPENDIX B

BRITISH PROPERTY FEDERATION
COLLATERAL WARRANTY FOR
FUNDING INSTITUTIONS
CoWa/F, SECOND EDITION 1990

N.B. All the Collateral Warranties given in this book are based upon this Form, and not the Third Edition reproduced in Appendix C.

Form of Agreement for

Collateral Warranty

for use where a warranty is to be given to a company providing finance for a proposed development

Pad of 5 sets of 4-page form

**Prepared and approved for use by
the British Property Federation
the Association of Consulting Engineers
the Royal Institute of British Architects and
the Royal Institution of Chartered Surveyors
after discussion with
the Association of British Insurers and other
organisations representing the major financial
institutions**

Form of Agreement for

Collateral Warranty for funding institutions | CoWa/F

The forms in this pad are for use where a warranty is to be given to a company providing finance for a proposed development. They must not in any circumstances be provided in favour of prospective purchasers or tenants.

General advice

1. The term 'collateral contract' or 'collateral warranty' is often used without due regard to the strict legal meaning of the phrase. It is used here for agreements with a funding institution putting up money for construction and development.

2. The purpose of the Agreement is to bind the party giving the warranty in contract where no contract would otherwise exist. This can have implications in terms of professional liability and could cause exposure to claims which might otherwise not have existed under common law.

3. The information and guidance contained in this note is designed to assist consultants confronted with a demand that collateral agreements be entered into.

4. The use of the word 'collateral' is not accidental. It is intended to refer to an agreement that is an adjunct to another or principal agreement, namely the conditions of appointment of the consultant. It is imperative therefore that before collateral warranties are executed the consultant's terms and conditions of appointment have been agreed between the client and the consultant and set down in writing.

5. The terms and conditions of the consultant's appointment may be 'under hand' or in the nature of a Deed and executed 'under seal'. In the latter case the length of time that claims may be brought under the agreement is extended from six years to twelve years.

6. This Form of Agreement for Collateral Warranty designed for use under hand or under seal. It should not be signed under seal when it is collateral to an appointment which is under hand.

7. The acceptance of a claim under the consultant' professional indemnity policy, brought under the of a collateral warranty, will depend on the terms and conditions of the policy in force the le when a claim is made.

8. Consultants with a current policy taken under the RIBA, RICSIS or ACE scheme will not have a claim refused simply on the basis at it ght under the terms of a collateral warranty ded at warranty is in this form. In other respects the will be treated in accordance with policy terms and conditions in the normal way. **Consultants insured under different policies** must seek the advice of their brokers or insurers.

9. **Amendment to the clauses should be resisted.** Insurers' approval as mentioned above is in respect of the unamended clauses only.

Published by
The British Property Federation Limited
35 Catherine Place
London SW1E 6DY
Telephone: 071-828 0111

© The British Property Federation, The Association of Consulting Engineers, The Royal Institute of British Architects and The Royal Institution of Chartered Surveyors. 1990.

ISBN 0 900101 08 6

Commentary on clauses

Recitals A, B and C are self-explanatory and need completion. The Consultant is described in the form as "The Firm". The following notes are to assist in understanding the use of the document:

Clause 1
This confirms the duty of care that will be owed and further that any obligation to the third party will be no greater than the obligations owed to the client. The terms "skill and care" or "skill, care and diligence" should accord with the conditions of engagem

Clause 2
As consultant it is give assurances beyond those to the effec t materials as listed will not be specified. Conc led u such materials by a contractor could possibl ccur hen the very careful restriction in terms of thi arti ar w rranty. Further materials may be added.

Clause
This o e con ultant to ensure that all fees due and owing a the warranty is entered into have been pa

Clause
This les the funding organisation to take over the sultar appointment from the client on terms that all re tstanding will be discharged by the funding authority (see use 7).

Clause 6
T affects the consultant's right to determine the ointment with the client in the sense that the funding authority will be given the opportunity of taking over the appointment, again subject to the payment of all fees which is the purpose of **Clause 7**.

Clause 8
This clause gives to the third party all the rights that they should reasonably expect in terms of use of drawings, etc. It does not give a licence to reproduce the designs contained within them other than for the purposes of the Development.

Clause 9
This is a provision confirming that professional indemnity insurance will be maintained in so far as it is reasonably possible to do so. Professional indemnity insurance is on the basis of annual contracts and the terms and conditions of a policy may change from renewal to renewal.

Clause 11
This clause indicates the right of assignment by the funding institution

Clause 12
This should be completed by identifying and agreeing the other parties required to sign similar agreements.

> N.B. The above advice and commentary is not intended to affect the interpretation of this Collateral Warranty. It is based on the terms of insurance current at the date of publication. All parties to the Agreement should ensure that terms of insurance have not changed.

Warranty Agreement   CoWa/F

Note

This form is to be used where the warranty is to be given to a company providing finance for the proposed development. Where that company is acting as an agent for a syndicate of banks, a recital should be added to refer to this as appropriate.

THIS AGREEMENT

is made the_____day of_____199_____

BETWEEN: –

(insert name of the Consultant)

(1) _____

of/whose registered office is situated at_____

_____ ("the Firm");

(insert name of the Firm's Client)

(2) _____

whose registered office is situated at_____

_____ ("the Client"); and

(insert name of the financier)

(3) _____

whose registered office is situated at_____

_____

("the Company" which term shall include all permitted assignees under this agreement)

WHEREAS: –

(insert description of the works)

A. The Company has entered into an agreement ("the Finance Agreement") with the Client for the provision of certain finance in connection with the carrying out of

_____

(insert address of the development)

_____

at_____

_____ ("the Development").

(insert date of appointment)

B. By a contract ("the Appointment") dated_____

(delete as appropriate)

the Client has appointed the Firm as [architects/consulting structural engineers/ consulting building services engineers/ surveyors] in connection with the Development.

(insert name of building contractor or "a building contractor to be selected by the Client")

C. The Client has entered or may enter into a building contract ("the Building Contract") with

_____

_____

for the construction of the Development.

CoWa/F   2nd Edition

Page 1

NOW IN CONSIDERATION OF THE PAYMENT OF ONE POUND (£1) BY THE COMPANY TO THE FIRM (RECEIPT OF WHICH THE FIRM ACKNOWLEDGES) IT IS HEREBY AGREED as follows: –

(to reflect
terms of the
Appointment)

1. The Firm warrants that it has exercised and will continue to exercise reasonable skill [and care] [care and diligence] in the performance of its duties to the Client under the Appointment, provided that the Firm shall have no greater liability to the Company by virtue of this Agreement than it would have had if the Company had been named as a joint client under the Appointment.

(Delete
where the
Firm is the
quantity
surveyor)

[2. Without prejudice to the generality of Clause 1, the Firm further warrants that it has exercised and will continue to exercise reasonable skill and care to see that, unless authorised by the Client in writing or, where such authorisation is given orally, confirmed by the Firm to the Client in writing, none of the following has been or will be specified by the Firm for use in the construction of those parts of the Development to which the Appointment relates: –

(a) high alumina cement in structural elements;

(b) wood wool slabs in permanent formwork to concrete;

(c) calcium chloride in admixtures for use in reinforced concrete;

(d) asbestos products;

(e) naturally occurring aggregates for use in reinforced concrete which do not comply with British Standard 882: 1983 and/or naturally occurring aggregates for use in concrete which do not comply with British Standard 8110: 1985.

(Further
specific
materials
may be added
by agreement)

(f)

]

3. The Company has no authority to issue any direction or instruction to the Firm in relation to performance of the Firm's duties under the Appointment unless and until the Company has given notice under Clauses 5 or 6.

4. The Firm acknowledges that the Client has paid all fees and expenses due and owing to the Firm under the Appointment up to the date of this Agreement. The Company has no liability to the Firm in respect of fees and expenses under the Appointment unless and until the Company has given notice under Clauses 5 or 6.

5. The Firm agrees that, in the event of the termination of the Finance Agreement by the Company, the Firm will, if so required by notice in writing given by the Company and subject to Clause 7, accept the instructions of the Company or its appointee to the exclusion of the Client in respect of the Development upon the terms and conditions of the Appointment. The Client acknowledges that the Firm shall be entitled to rely on a notice given to the Firm by the Company under this Clause 5 as conclusive evidence for the purposes of this Agreement of the termination of the Finance Agreement by the Company.

6. The Firm further agrees that it will not without first giving the Company not less than twenty one days' notice in writing exercise any right it may have to terminate the Appointment or to treat the same as having been repudiated by the Client or to discontinue the performance of any duties to be performed by the Firm pursuant thereto. The Firm's right to terminate the Appointment with the Client or treat the same as having been repudiated or discontinue performance shall cease if, within such period of notice and subject to Clause 7, the Company shall give notice in writing to the Firm requiring the Firm to accept the instructions of the Company or its appointee to the exclusion of the Client in respect of the Development upon the terms and conditions of the Appointment.

7. It shall be a condition of any notice given by the Company under Clauses 5 or 6 that the Company or its appointee accepts liability for payment of the fees payable to the Firm under the Appointment and for performance of the Client's obligations under the Appointment including payment of any fees outstanding at the date of such notice. Upon the issue of any notice by the Company under Clauses 5 or 6, the Appointment shall continue in full force and effect as if no right of termination on the part of the Firm had arisen and the Firm shall be liable to the Company or its appointee under the Appointment in lieu of its liability to the Client. If any notice given by the Company under Clauses 5 or 6 requires the Firm to accept the instructions of the Company's appointee, the Company shall be liable to the Firm as guarantor for the payment of all sums from time to time due to the Firm from the Company's appointee.

8. The copyright in all drawings, reports, specifications, bills of quantities, calculations and other similar documents provided by the Firm in connection with the Development shall remain vested in the Firm but the Company and its appointee shall have a licence to copy and use such drawings and other documents and to reproduce the designs contained in them for any purpose related to the Development including, but without limitation, the construction, completion, maintenance, letting, promotion, advertisement, reinstatement and repair of the Development. The Company and its appointee shall have a licence to copy and use such drawings and other documents for the construction of the Development but such use shall not include a licence to reproduce the designs contained in them for any extension of the Development. The Firm shall not be liable for any such use by the Company or its appointee of any drawings and other documents for any purpose other than that for which the same were prepared and provided by the Firm.

(insert **9.** amount)
(insert period)
The Firm shall maintain professional indemnity insurance in an amount of not less than _____ pounds (£) for any one occurrence or series of occurrences arising out of any one event for a period of ____ years from the date of practical completion of the Development for the purposes of the Building Contract, provided always that such insurance is available at commercially reasonable rates. The Firm shall immediately inform the Company if such insurance ceases to be available at commercially reasonable rates in order that the Firm and the Company can discuss means of best protecting the respective positions of the Company and the Firm in respect of the Development in the absence of such insurance. As and when it is reasonably required to do so by the Company or its appointee under Clauses 5 or 6, the Firm shall produce for inspection documentary evidence that its professional indemnity insurance is being maintained.

10. The Client has agreed to be a party to this Agreement for the purposes of Clause 12 and for acknowledging that the Firm shall not be in breach of the Appointment by complying with the obligations imposed on it by Clauses 5 and 6.

11. This Agreement may be assigned by the Company by way of absolute legal assignment to another company providing finance or re-finance in connection with the Development without the consent of the Client or the Firm being required and such assignment shall be effective upon written notice thereof being given to the Client and to the Firm.

(insert names and/ **12.** or descriptions of other parties required to sign warranty agreements)
The Client undertakes to the Firm that warranty agreements in the Model Form CoWa/F published by the British Property Federation or in substantially similar form have been or will be entered into between_____

_____

_____

on the one hand and the Company on the other hand.

13. Any notice to be given by the Firm hereunder shall be deemed to be duly given if it is delivered by hand at or sent by registered post or recorded delivery to the Company at its registered office and any notice to be given by the Company hereunder shall be deemed to be duly given if it is addressed to "The Senior Partner"/"the Managing Director" and delivered by hand at or sent by registered post or recorded delivery to the above-mentioned address of the Firm or to the principal business address of the Firm for the time being and, in the case of any such notices, the same shall if sent by registered post or recorded delivery be deemed to have been received forty eight hours after being posted.

(Alternatives: delete as appropriate)

AS WITNESS the hands of the parties the day and year first before written.

(These must only apply if the Appointment is under seal)

IN WITNESS whereof the partners in the Firm have hereunto set their hands and seals and the Common Seals of the Company and the C___ are hereunto affixed the day and year first before written. ·

IN WITNESS ___ the common seals of the parties were hereunto affixed the day and year first before written.

(NB: All parties referred to in Clause 12 must use the same attestation clause)

SPECIMEN

APPENDIX C

BRITISH PROPERTY FEDERATION
COLLATERAL WARRANTY FOR
FUNDING INSTITUTIONS
CoWa/F, THIRD EDITION 1992

N.B. All the Collateral Warranties given in this book are based upon the Second Edition, reproduced in Appendix B, and not upon this Edition.

SPECIMEN

CoWa/F

Third Edition (1992)

Form of Agreement for

Collateral Warranty

for use where a warranty is to be given
to a company providing finance
for a proposed development

Pad of 5 sets of 5-page form.

Prepared and approved for use by
the British Property Federation
the Association of Consulting Engineers
the Royal Incorporation of Architects in Scotland
the Royal Institute of British Architects and
the Royal Institution of Chartered Surveyors
after discussion with
the Association of British Insurers and other
organisations representing the major financial
institutions

Form of Agreement for

Collateral Warranty for funding institutions

CoWa/F

The forms in this pad are for use where a warranty is to be given to a company providing finance for a proposed development. They must not in any circumstances be provided in favour of prospective purchasers or tenants.

General advice

1. The term "collateral agreement", "duty of care letter" or "collateral warranty" is often used without due regard to the strict legal meaning of the phrase. It is used here for agreements with a funding institution putting up money for construction and development.

2. The purpose of the Agreement is to bind the party giving the warranty in contract where no contract would otherwise exist. This can have implications in terms of professional liability and could cause exposure to claims which might otherwise not have existed under Common Law.

3. The information and guidance contained in this note is designed to assist consultants faced with a request that collateral agreements be entered into.

4. The use of the word "collateral" is not accidental. It is intended to refer to an agreement that is an adjunct to another or principal agreement, namely the conditions of appointment of the consultant. It is imperative therefore that before collateral warranties are executed the consultant's terms and conditions of appointment have been agreed between the client and the consultant and set down in writing.

5. Under English Law the terms and conditions of the consultant's appointment may be "under hand" or executed as a Deed. In the latter case the length of time that claims may be brought under the Agreement is extended from six years to twelve years.

6. Under English Law this Form of Agreement for Collateral Warranty is designed for use under hand or to be executed as a Deed. It should not be signed as a Deed when it is collateral to an appointment which is under hand.

7. The acceptance of a claim under the consultant's professional indemnity policy, brought under the terms of a collateral warranty, will depend upon the terms and conditions of the policy in force at the time when a claim is made.

8. Consultants with a current indemnity insurance policy taken out under the RIBA, RICSIS, ASE or H___ schemes will not have a claim refused simply on the basis that it is brought under the terms of a collateral warranty provided that warranty is in this ___. In other respects the claim will be treated in accordance with policy terms and conditions in the normal way. **Consultants insured under different policies** must seek the advice of their brokers or insurers.

9. **Amendment to the clauses should be resisted.** Insurers' approval as mentioned a___ in respect of the unamended clauses only.

Commentary on Clauses

Recitals A, B and C are self-explanatory and need completion. The Consultant is described in the form as "The Firm". The following notes are to assist in understanding the use of the document:

Clause 1
This confirms the duty of care that will be owed to the Company. The words in square brackets enable the clause to reflect exactly the provision contained within the terms and conditions of the Appointment.

Paragraphs (a) and (b) qualify and limit in two ways the Firm's liability in the event of a breach of the duty of care.

1 (a)   By this provision the Firm's potential liability is limited. intention is that the effect of "several" liability at Common Law is negated. When the Firm agrees - probably at the time of appointment - to sign a warranty at a future date, the list should include the parties, ___ known otherwise the description or profession, of those responsible for the design of the relevant parts of the Development and the general ___ top ___. When the warranty is signed, the list should be completed with the names of those previously referred to by description or profession.

1 (b)   By this clause, the Company is bound by any limitations on liability that may exist in the conditions of the Appointment. Furthermore, the consultant has the same rights of defence that would have been available had the relevant claim been made by the client under the Appointment.

Clause 2
As a consultant it is not possible to ___ and ___ beyond those to the effect that materials as listed have not or will be specified. Concealed use of such materials by a contractor could possibly occur, hence the very careful restriction in terms of this particular warranty. Further materials may be added.

Clause 4
This obliges the consultant to ensure that all fees due and owing including VAT at the time the warranty is entered into have been paid.

Clause 5
This entitles the funding organistion to take over the consultant's appointment from the client on terms that all fees outstanding will be discharged by the funding authority (see Clause 7).

Clause 6
This affects the consultant's right to determine the appointment with the client in the sense that the funding authority will be given the opportunity of taking over the appointment, again subject to the payment of all fees which is the purpose of **Clause 7**.

___ use by the Company of drawings and associated documents ___ssary in most cases. By this clause, the Company is given the rights that ___ be reasonably expected but it does not allow the reproduction of the designs for any purpose outside the scope of the Development.

Clause 9
This confirms that professional indemnity insurance will be maintained in so far as it is reasonably possible to do so. Professional indemnity insurance is on the basis of annual contracts and the terms and conditions of a policy may change from renewal to renewal.

Clause 11
This clause indicates the right of assignment by the funding institution.

Clause 11S
This is applicable in Scotland in relation to assignations.

Clause 12
This identifies the method of giving Notice under Clauses 5, 6, 11 &11S

Clause 13
This needs completion. The clause makes clear that any liability that the Firm has by virtue of this Warranty ceases on the expiry of the stated period of years after practical completion of the Premises. (Note: the practical completion of the Development may be later).

Under English law the period should not exceed 6 years for agreements under hand, nor 12 years for those executed as a Deed.

In Scotland, the Prescription and Limitations (Scotland) Act 1973 prescribes a 5 year period.

Clause 14 and Attestation below
The appropriate method of execution by the Firm, the Client and the Company should be checked carefully.

Clause 14S and Testing Clause below
This assumes the Firm is a partnership and the Client and the Company are Limited Companies. Otherwise legal advice should be taken.

Published by
The British Property Federation Limited
35 Catherine Place, London SW1E 6DY Telephone: 071-828 0111

© The British Property Federation, The Association of Consulting Engineers, The Royal Incorporation of Architects in Scotland, The Royal Institute of British Architects and The Royal Institution of Chartered Surveyors. 1992.

ISBN 0 900101 08 6

N.B. The above advice and commentary is not intended to affect the interpretation of this Collateral Warranty. It is based on the terms of insurance current at the date of publication. All parties to the Agreement should ensure the terms of insurance have not changed.

Warranty Agreement   CoWa/F

Note

This form is to be used where the warranty is to be given to a company providing finance for the proposed development. Where that company is acting as an agent for a syndicate of banks, a recital should be added to refer to this as appropriate.

THIS AGREEMENT

(In Scotland, leave blank. For applicable date see Testing Clause on page 5)

is made the .... day of ..../......................... 19 ..../..............

BETWEEN:-

(insert name of the Consultant)

(1) ....

of/whose registered office is situated at ....

.... ("the Firm");

(insert name of the Firm's Client)

(2) ....

whose registered office is situated at ....

.... ("the Client"); and

(insert name of the financier)

(3) ....

whose registered office is situated at ....

("the Company" which term shall include all permitted assignees under this agreement).

WHEREAS:-

A. The Company has entered into an agreement ("the Finance Agreement") with the Client for the provision of certain finance in connection with the carrying out of

(insert description of the works)

....

....

(insert address of the development)

at...

....

.... ("the Development").

(insert date of appointment) (delete/complete as appropriate)

B. By a contract ("the Appointment") dated ....
the Client has appointed the Firm as [architects/consulting structural engineers/consulting building services engineers/   surveyors] in connection with the Development.

(insert name of building contractor or "a building contractor to be selected by the Client")

C. The Client has entered or may enter into a building contract ("the Building Contract") with

....

....

....

for the construction of the Development.

NOW IN CONSIDERATION OF THE PAYMENT OF ONE POUND (£1) BY THE COMPANY TO THE FIRM (RECEIPT OF WHICH THE FIRM ACKNOWLEDGES) IT IS HEREBY AGREED as follows:-

(delete "and care" or "care and diligence" to reflect terms of the Appointment)

1. The Firm warrants that it has exercised and will continue to exercise reasonable skill [and care] [care and diligence] in the performance of its duties to the Client under the Appointment. In the event of any breach of this warranty:

    (a) the Firm's liability for costs under this Agreement shall be limited to that proportion of the Company's losses which it would be just and equitable to require the Firm to pay having regard to the extent of the Firm's responsibility for the same and on the basis that ......

(insert the names of other intended warrantors)

    ....
    ....
    ....
    ....
    .... shall be deemed to have provided contractual undertakings on terms no less onerous than this Clause 1 to the Company in respect of the performance of their services in connection with the Development and shall be deemed to have paid to the Company such proportion which it would be just and equitable for them to pay having regard to the extent of their responsibility;

    (b) the Firm shall be entitled in any action or proceedings by the Company to rely on any limitation in the Appointment and to raise the equivalent rights in defence of liability as it would have against the Client under the Appointment;

(delete where the Firm is the quantity surveyor)

2. [Without prejudice to the generality of Clause 1, the Firm further warrants that it has exercised and will continue to exercise reasonable skill and care to see that, unless authorised by the Client in writing or, where such authorisation is given orally, confirmed by the Firm to the Client in writing, none of the following has been or will be specified by the Firm for use in the construction of those parts of the Development to which the Appointment relates:-

    (a) high alumina cement in structural elements;

    (b) wood wool slabs in permanent formwork to concrete;

    (c) calcium chloride in admixtures for use in reinforced concrete;

    (d) asbestos products;

    (e) naturally occurring aggregates for use in reinforced concrete which do not comply with British Standard 882: 1983 and/or naturally occurring aggregates for use in concrete which do not comply with British Standard 8110: 1985.

(further specific materials may be added by agreement)

    (f)

    ]

3. The Company has no authority to issue any direction or instruction to the Firm in relation to performance of the Firm's services under the Appointment unless and until the Company has given notice under Clauses 5 or 6.

4. The Firm acknowledges that the Client has paid all fees and expenses properly due and owing to the Firm under the Appointment up to the date of this Agreement. The Company has no liability to the Firm in respect of fees and expenses under the Appointment unless and until the Company has given notice under Clauses 5 or 6.

5. The Firm agrees that, in the event of the termination of the Finance Agreement by the Company, the Firm will, if so required by notice in writing given by the Company and subject to Clause 7, accept the instructions of the Company or its appointee to the exclusion of the Client in respect of the Development upon the terms and conditions of the Appointment. The Client acknowledges that the Firm shall be entitled to rely on a notice given to the Firm by the Company under this Clause 5 as conclusive evidence for the purposes of this Agreement of the termination of the Finance Agreement by the Company.

6. The Firm further agrees that it will not without first giving the Company not less than twenty one days' notice in writing exercise any right it may have to terminate the Appointment or to treat the same as having been repudiated by the Client or to discontinue the performance of any services to be performed by the Firm pursuant thereto. Such right to terminate the Appointment with the Client or treat the same as having been repudiated or discontinue performance shall cease if, within such period of notice and subject to Clause 7, the Company shall give notice in writing to the Firm requiring the Firm to accept the instructions of the Company or its appointee to the exclusion of the Client in respect of the Development on the terms and conditions of the Appointment.

7. It shall be a condition of any notice given by the Company under Clauses 5 or 6 that the Company or its appointee accepts liability for payment of the fees and expenses payable to the Firm under the Appointment and for performance of the Client's obligations including payment of any fees and expenses outstanding at the date of such notice. Upon the issue of any notice by the Company under Clauses 5 or 6 the Appointment shall continue in full force and effect as if no right of termination on the part of the Firm had arisen and the Firm shall be liable to the Company and its appointee under the Appointment in lieu of its liability to the Client. If any notice given by the Company under Clauses 5 or 6 requires the Firm to accept the instructions of the Company's appointee, the Company shall be liable to the Firm as guarantor for the payment of all sums from time to time due to the Firm from the Company's appointee.

8. The copyright in all drawings, reports, models, specifications, bills of quantities, calculations and other similar documents provided by the Firm in connection with the Development (together referred to in this Clause 8 as "the Documents") shall remain vested in the Firm but, subject to the Firm having received payment of any fees agreed as properly due under the Appointment, the Company and its appointee shall have a licence to copy and use the Documents and to reproduce the designs and content of them for any purpose related to the Premises including, but without limitation, the construction, completion, maintenance, letting, promotion, advertisement, reinstatement, refurbishment and repair of the Development. Such licence shall enable the Company and its appointee to copy and use the Documents for the extension of the Development but such use shall not include a licence to reproduce the designs contained in them for any extension of the Development. The Firm shall not be liable for any such use by the Company or its appointee of any of the Documents for any purpose other than that for which the same were prepared by or on behalf of the Firm.

(insert amount)

(insert period)

9. The Firm shall maintain professional indemnity insurance in an amount of not less than _____ pounds (£   ) for any one occurrence or series of occurrences arising out of any one event for a period of   years from the date of practical completion of the Development for the purposes of the Building Contract, provided always that such insurance is available at commercially reasonable rates. The Firm shall immediately inform the Company if such insurance ceases to be available at commercially reasonable rates in order that the Firm and the Company can discuss means of best protecting the respective positions of the Company and the Firm in respect of the Development in the absence of such insurance. As and when it is reasonably requested to do so by the Company or its appointee under the Clauses 5 or 6, the Firm shall produce for inspection documentary evidence that its professional indemnity insurance is being maintained.

10. The Client has agreed to be a party to this Agreement for the purposes of acknowledging that the Firm shall not be in breach of the Appointment by complying with the obligations imposed on it by Clauses 5 and 6.

(delete if under Scots law) [11. This Agreement may be assigned by the Company by way of absolute legal assignment to another company providing finance or re-finance in connection with the carrying out of the Development without the consent of the Client or the Firm being required and such assignment shall be effective upon written notice thereof being given to the Client and to the Firm.]

(delete if under English law) [11S. *The Company shall be entitled to assign or transfer its rights under this Agreement to any other company providing finance or re-finance in connection with the carrying out of the Development without the consent of the Client or the Firm being required subject to written notice of such assignation being given to the Firm in accordance with Clause 12 hereof.*]

12. Any notice to be given by the Firm hereunder shall be deemed to be duly given if it is delivered by hand at or sent by registered post or recorded delivery to the Company at its registered office and any notice given by the Company hereunder shall be deemed to be duly given if it is addressed to "The Senior Partner"/"The Managing Director" and delivered by hand at or sent by registered post or recorded delivery to the above-mentioned address of the Firm or to the principal business address of the Firm for the time being and, in the case of any such notices, the same shall if sent by registered post or recorded delivery be deemed to have been received forty eight hours after being posted.

(complete as appropriate) 13. No action or proceedings for any breach of this Agreement shall be commenced against the Firm after the expiry of _____ years from the date of practical completion of the Premises under the Building Contract.

(delete if under Scots law) [14. The construction validity and performance of this Agreement shall be governed by English Law and the parties agree to submit to the non-exclusive jurisdiction of the English Courts.

(alternatives: delete as appropriate) [**AS WITNESS** the hands of the parties the day and year first before written.

(for Agreement executed under hand and NOT as a Deed) Signed by or on behalf of the Firm ....

in the presence of: ....

Signed by or on behalf of the Client ....

in the presence of: ....

Signed by or on behalf of the Company ....

in the presence of: ....]

(this must only apply if the Appointment is executed as a Deed) [**IN WITNESS WHEREOF** this Agreement was executed as a Deed and delivered the day and year first before written.

by the Firm

....

....

....

by the Client

....

....

....

by the Company

....

....

....]]

CoWa/F 3rd Edition
© BPF, ACE, RIAS, RIBA, RICS 1992

Page 4

(delete if under English law) [14S. *This Agreement shall be construed and the rights of the parties and all matters arising hereunder shall be determined in all respects according to the Law of Scotland.*

IN WITNESS WHEREOF *these presents are executed as follows:-*

SIGNED by the above named Firm at ....

on the .... *day of* .... *Nineteen hundred and*..............................

as follows:-

....(Firm's signature)

Signature .... *Full Name* ....

Address ....

....*Occupation* ....

Signature .... *Full Name* ....

Address ....

....*Occupation* ....

SIGNED by the above named Client at ....

on the .... *day of* .... *Nineteen hundred and*..............................

as follows:-

For and on behalf of the Client

....*Director/Authorised Signatory*

....*Director/Authorised Signatory*

SIGNED by the above named Company at ....

on the .... *day of* .... *Nineteen hundred and*..............................

as follows:-

For and on behalf of the Company

....*Director/Authorised Signatory*

....*Director/Authorised Signatory*]

APPENDIX D

BRITISH PROPERTY FEDERATION
COLLATERAL WARRANTY FOR
PURCHASERS AND TENANTS
CoWa/P&T, 1992

N.B. All the Collateral Warranties given in this book are based upon the Form reproduced in Appendix B, and not upon this Form.

SPECIMEN

Form of Agreement for

Collateral
Warranty

**for use where a warranty is to be given to a
purchaser or tenant of premises
in a commercial and/or industrial development**

Pad of 5 sets of 4-page form

**Prepared and approved for use by
the British Property Federation
the Association of Consulting Engineers
the Royal Incorporation of Architects in Scotland
the Royal Institute of British Architects and
the Royal Institution of Chartered Surveyors**

SPECIMEN

Form of Agreement for

Collateral Warranty
for purchasers & tenants | CoWa/P&T

The forms in this pad are for use where a warranty is to be given to a purchaser or tenant of a whole building in a commercial and/or industrial development, or a part of such a building. It is essential that the number of warranties to be given to tenants in one building should sensibly be limited.

General advice

1. The term "collateral agreement", "duty of care letter" or "collateral warranty" is often used without due regard to the strict legal meaning of the phrase. It is used here for agreements with tenants or purchasers of the whole or part of a commercial and/or industrial development.

2. The purpose of the Agreement is to bind the party giving the warranty in contract where no contract would otherwise exist. This can have implications in terms of professional liability and could cause exposure to claims which might otherwise not have existed under Common Law.

3. The information and guidance contained in this note is designed to assist consultants faced with a request that collateral agreements be entered into.

4. The use of the word 'collateral' is not accidental. It is intended to refer to an agreement that is an adjunct to another or principal agreement, namely the conditions of appointment of the consultant. It is imperative therefore that before collateral warranties are executed the consultant's terms and conditions of appointment have been agreed between the client and the consultant and set down in writing.

5. Under English Law the terms and conditions of the consultant's appointment may be 'under hand' or executed as a Deed. In the latter case the length of time that claims may be brought under the Agreement is extended from six years to twelve years.

6. Under English Law this Form of Agreement for Collateral Warranty is designed for use under hand or to be executed as a Deed. It should not be signed as a Deed when it is collateral to an appointment which is under hand.

7. The acceptance of a claim under the consultant's professional indemnity policy, brought under the terms of a collateral warranty, will depend upon the terms and conditions of the policy in force at the time when a claim is made.

8. Consultants with a current indemnity insurance policy taken out under the RIBA, RICSIS, ACE or RIAS schemes will not have a claim refused simply on the basis that it is brought under the terms of a collateral warranty provided that warranty is in this form. In other respects the claim will be treated in accordance with policy terms and conditions in the normal way. **Consultants insured under different policies** must seek the advice of their brokers or insurers.

9. **Amendment to the clauses should be resisted.** Insurers' approval as mentioned above is in respect of the unamended clauses only.

Commentary on Clauses

Recital A.

This needs completion.

When this warranty is to be given in favour of a purchaser or tenant of part of the Development, the following words in square brackets must be deleted.

["The Premises" are also referred to as "the Development" in this Agreement.]

Care must be taken in describing "the Premises" accurately.

When this warranty is to be given in favour of a purchaser or tenant of the entire development, the terms "the Premises" and "the Development" are synonymous.

The following words in square brackets must be deleted

[forming part of. ....

at. ....("the Development").]

Recitals B & C

These are self explanatory but need completion.

Clause 1

This confirms the duty of care that will be owed to the Purchaser/the Tenant. The words in square brackets enable the clause to reflect correctly the provisions contained within the terms and conditions of the Appointment.

Paragraphs (a),(b) and (c) qualify and limit in three ways the Firm's liability in the event of a breach of the duty of care.

1 (a)   By this provision, the Firm is liable for the reasonable costs of repair renewal and or reinstatement of the Development insofar as the Purchaser/the Tenant has a financial obligation to pay or contribute to the cost of that repair. Other losses are expressly excluded.

1 (b)   By this provision the Firm's potential liability is limited. The intention is that the effect of "several" liability at Common Law is negated. When the Firm agrees – probably at the time of appointment – to sign a warranty at a future date, the list should include the names, if known, or otherwise the description or profession, of those responsible for the design of the relevant parts of the Development and the general contractor. When the warranty is signed, the list should be completed with the names of those previously referred to by description or profession.

1 (c)   By this clause, the Purchaser/ the Tenant is bound by any limitations on liability that may exist in the conditions of the Appointment. Furthermore, the consultant has the same rights of defence that would have been available had the relevant claim been made by the Client under the Appointment.

1 (d)   This states the relationship between the Firm and any consultant employed by the Purchaser/the Tenant to survey the premises.

Clause 2

As a consultant it is not possible to give assurances beyond those to the effect that materials as listed have not been nor will be specified. Concealed use of such materials by a contractor could possibly occur, hence the very careful restriction in terms of this particular warranty. Further materials may be added.

N.B.   The above advice and commentary is not intended to affect the interpretation of this Collateral Warranty. It is based on the terms of insurance current at the date of publication. All parties to the Agreement should ensure the terms of insurance have not changed.

Clause 3

This obliges the consultant to ensure that all fees due and owing including VAT at the time the warranty is entered into have been paid.

Clause 4

This is designed to make it clear that the Purchaser/the Tenant has no power or authority to direct or instruct the Firm in its duties to the Client.

Clause 5

Reasonable use by the Purchaser/the Tenant of drawings and associated documents is necessary in most cases. By this clause, the Purchaser/ the Tenant is given the rights that might be reasonably expected but it does not allow the reproduction of the designs for any purpose outside the scope of the Development.

Clause 6

This confirms that professional indemnity insurance will be maintained in so far as it is reasonably possible to do so. Professional indemnity insurance is on the basis of annual contracts and the terms and conditions of a policy may change from renewal to renewal.

Clause 7

This allows the Purchaser/the Tenant to assign the benefit of this Warranty provided it is done by formal legal assignment and relates to the entire interest of the original Purchaser/Tenant. By this clause any right of assignment may be limited or extinguished. If it is to be extinguished the word "not" shall be inserted after "may" and all words after "the Purchaser/ the Tenant" deleted. If it is agreed that there should be a limited number of assignments, the precise number should be inserted in the space between "assigned" and "by the Purchaser/the Tenant".

Clause 7S

This is applicable in Scotland in relation to assignations. Completion is as for Clause 7.

Clause 8

This identifies the method of giving Notice under Clause 7 & 7S.

Clause 9

This needs completion. The clause makes clear that any liability that the Firm has by virtue of this Warranty ceases on the expiry of the stated period of years after practical completion of the Premises. (Note: the practical completion of the Development may be later).

Under English law the period should not exceed 6 years for agreements under hand, nor 12 years for those executed as a Deed.

In Scotland, the Prescription and Limitations (Scotland) Act 1973 prescribes a 5 year period.

Clause 10 and Attestation below

The appropriate method of execution by the Firm and the Purchaser/the Tenant should be checked carefully.

Clause 10S and Testing Clause below

This assumes the Firm is a partnership and the Purchaser/the Tenant is a Limited Company. Otherwise legal advice should be taken.

Published by
The British Property Federation Limited
35 Catherine Place, London SW1E 6DY Telephone: 071-828 0111

© The British Property Federation, The Association of Consulting Engineers, The Royal Incorporation of Architects in Scotland, The Royal Institute of British Architects and The Royal Institution of Chartered Surveyors. 1992.

ISBN 0 900101 08 7

British Property Federation

35 Catherine Place, London SW1E 6DY
Telephone: 071-828 0111
Facsimile: 071-834 3442

MODEL FORM OF COLLATERAL WARRANTY
FOR PURCHASERS AND TENANTS – CoWa/P&T

IMPORTANT ADDITIONAL NOTES FOR CLIENTS

Clients will generally wish to include in their Conditions of Engagement for Consultants, a clause to the effect that the consultant shall be prepared to enter into a stated number of collateral warranties in favour of possible purchasers, a number of tenants in a multi-occupied building and possibly to a funding institution. Model forms of the collateral warranty be used should be attached to the conditions of engagement, and should be completed so that the consultant is aware of his obligations and liabilities at the time he quotes his fee for a project. CoWa/F should be used for funding institutions as a model.

This model form of collateral warranty for purchasers and tenants, "CoWa/P&T", has been agreed by the BPF, the ACE, the RIAS, the RIBA and the RICS after consultation with the Association of British Insurers. It will be acceptable to many purchasers and tenants. It should be noted that in its unamended form it is acceptable to certain insurers of the three professional organisations. However, some purchasers and tenants may demand additional features. These are listed below but it should be remembered that insistence upon them may negate the consultant's professional indemnity insurance.

In this model form for purchasers and tenants, CoWa/P&T, the following points should be noted in addition to the deletions and additions in the printed guidance notes.

1.   Economic and Consequential Loss

Clause 1(a) limits the consultant's liability to the recovery of costs of repair, renewal and/or reinstatement of any part or parts of the Development if the consultant has been in breach of the warranty, i.e. negligent. The clause continues by saying that the consultant shall **not** be liable for any other losses. In other words, if the defect caused by the consultant's negligence causes consequential loss such as loss of profit, loss of production, the cost of removal to, and the renting of, alternative premises etc. then the consultant's liability for them is **excluded.**

Some purchasers and tenants will wish to hold the consultants responsible for "consequential loss". Some consultants may be able to extend their professional indemnity cover to include "consequential loss" and some may be able to obtain additional but separate cover for these losses. In both cases, however, there will almost certainly be a limit to the extent of the consultant's liability. Those clients who wish to extend the consultant's responsibility to cover economic and consequential loss – and who can persuade the consultants to provide adequate insurance cover – should **delete** the last sentence of Clause 1a which reads:

"The Firm shall not be liable for other losses incurred by the Purchaser/the Tenant".

British Property Federation Limited
Registration No. 778293 England. Registered office, as above

Over . . . /

357

The following sentence should be **inserted** in lieu:

> "The Firm shall in addition be liable for other losses incurred by the Purchaser/the Tenant provided that such additional liability of the Firm shall not exceed £..................................... in respect of each breach of the Firm's warranty contained in this Clause 1".

The figure to be inserted as the limit is often the same as the consultants' professional indemnity cover.

2.   **Assignment** – Clause 7

Purchasers and Tenants will wish to have the facility to assign their collateral warranties when selling their property or assigning their leases. On the other hand insurers will wish to limit the number of assignments because this limits their liability. Clients must give careful consideration to completion of Clause 7.

Consultants who wish to deny any assignment will attempt to insert "not" between "This Agreement may" and "be assigned", in line 1 of Clause 7. This would be   **unacceptable** to most purchasers and tenants. Clients are therefore advised to draw a line between "may" and "be assigned".

There is a further space in line 1 of Clause 7 between "be assigned" and "by the Purchaser/the Tenant". This enables the client to insert the number of assignments which may be allowed by the consultant. Ownership of premises does not change frequently, nor are leases often assigned. A reasonable number inserted in this space would meet most requirements of purchasers or tenants.

3.   **Limitation of liability**

Clause 9 removes all doubt about the period of liability under the agreement. "6 years" should be inserted for agreement under hand and "12 years" if the original appointment is executed as a Deed and the agreement is also to be executed as a Deed. Requests from consultants to include shorter periods should be resisted.

21st February 1992
The British Property Federation

Warranty Agreement   CoWa/P&T

THIS AGREEMENT

is made the ....day of 199

BETWEEN:-

(insert name of the Consultant)

(1) ....

of/whose registered office is situated at ....

.... ("the Firm"), and

(insert name of the Purchaser/the Tenant)

(2) ....

whose registered office is situated at ....

....

(delete as appropriate)

("the Purchaser"/"the Tenant" which term shall include all permitted assignees under this Agreement).

WHEREAS:-

(delete as appropriate)

A.   The Purchaser/the Tenant has entered into an agreement to purchase/an agreement to lease/a lease with

....

.... ("the Client") relating to

(insert description of the premises)

....

....

.... ("the Premises")

(delete as appropriate)

[forming part of ....

(insert description of the development)

....

(insert address of the development)

at ....

.... ("the Development").]

(delete as appropriate)

["The Premises" are also referred to as "the Development" in this Agreement.]

(insert date of appointment)
(delete/complete as appropriate)

B.   By a contract ("the Appointment") dated ....
the Client has appointed the Firm as [architects/consulting structural engineers/consulting building services engineers/   surveyors] in connection with the Development.

C.   The Client has entered or may enter into a contract ("the Building Contract") with

(insert name of building contractor or "a building contractor to be selected by the Client")

....

....

....

for the construction of the Development.

NOW IN CONSIDERATION OF THE PAYMENT OF ONE POUND (£1) BY THE PURCHASER/ THE TENANT TO THE FIRM (RECEIPT OF WHICH THE FIRM ACKNOWLEDGES) IT IS HEREBY AGREED as follows:-

1. The Firm warrants that it has exercised and will continue to exercise reasonable skill [and care] [care and diligence] in the performance of its services to the Client under the Appointment. In the event of any breach of this warranty:

    (a) subject to paragraphs (b) and (c) of this clause, the Firm shall be liable for the reasonable costs of repair renewal and/or reinstatement of any part or parts of the Development to the extent that the Purchaser/the Tenant is liable either directly or by way of financial contribution for the same. The Firm shall not be liable for other losses incurred by the Purchaser/the Tenant;

    (b) the Firm's liability for costs under this Agreement shall be limited to that proportion of such costs which it would be just and equitable to require the Firm to pay having regard to the extent of the Firm's responsibility for the same and on the basis that ....

    ....
    ....
    ....
    ....
    ....shall be deemed to have provided contractual undertakings on terms no less onerous than this Clause 1 to the Purchaser/the Tenant in respect of the performance of their services in connection with the Development and shall be deemed to have paid to the Purchaser/the Tenant such proportion which it would be just and equitable for them to pay having regard to the extent of their responsibility;

    (c) the Firm shall be entitled in any action or proceedings by the Purchaser/the Tenant to rely on any limitation in the Appointment and to raise the equivalent rights in defence of liability as it would have against the Client under the Appointment;

    (d) the obligations of the Firm under or pursuant to this Clause 1 shall not be released or diminished by the appointment of any person by the Purchaser/the Tenant to carry out any independent enquiry into any relevant matter.

2. [Without prejudice to the generality of Clause 1, the Firm further warrants that it has exercised and will continue to exercise reasonable skill and care to see that, unless authorised by the Client in writing or, where such authorisation is given orally, confirmed by the Firm to the Client in writing, none of the following has been or will be specified by the Firm for use in the construction of those parts of the Development to which the Appointment relates:-

    (a) high alumina cement in structural elements;

    (b) wood wool slabs in permanent formwork to concrete;

    (c) calcium chloride in admixtures for use in reinforced concrete;

    (d) asbestos products;

    (e) naturally occurring aggregates for use in reinforced concrete which do not comply with British Standard 882: 1983 and/or naturally occurring aggregates for use in concrete which do not comply with British Standard 8110: 1985.

    (f)

]

3. The Firm acknowledges that the Client has paid all fees and expenses properly due and owing to the Firm under the Appointment up to the date of this Agreement.

4. The Purchaser/the Tenant has no authority to issue any direction or instruction to the Firm in relation to the Appointment.

5. The copyright in all drawings, reports, models, specifications, bills of quantities, calculations and other documents and information prepared by or on behalf of the Firm in connection with the Development (together referred to in this Clause 5 as "the Documents") shall remain vested in the Firm but, subject to the Firm having received payment of any fees agreed as properly due under the Appointment, the Purchaser/the Tenant and its appointee shall have a licence to copy and use the Documents and to reproduce the designs and content of them for any purpose related to the Premises including, but without limitation, the construction, completion, maintenance, letting, promotion, advertisement, reinstatement, refurbishment and repair of the Premises. Such licence shall enable the Purchaser/the Tenant and its appointee to copy and use the Documents for the extension of the Premises but such use shall not include a licence to reproduce the designs contained in them for any extension of the Premises. The Firm shall not be liable for any use by the Purchaser/the Tenant or its appointee of any of the Documents for any purpose other than that for which the same are prepared by or on behalf of the Firm.

(insert amount)

(insert period)

6. The Firm shall maintain professional indemnity insurance in an amount of not less than pounds (£   ) for any one occurrence or series of occurrences arising out of any one event for a period of   years from the date of practical completion of the Premises under the Building Contract, provided always that such insurance is available at commercially reasonable rates. The Firm shall immediately inform the Purchaser/the Tenant if such insurance ceases to be available at commercially reasonable rates in order that the Firm and the Purchaser/the Tenant can discuss means of best protecting the respective positions of the Purchaser/the Tenant and the Firm in the absence of such insurance. As and when it is reasonably requested to do so by the Purchaser/the Tenant or its appointee the Firm shall produce for inspection documentary evidence that its professional indemnity insurance is being maintained.

(insert number of times)

(delete if under Scots law)

[7. This Agreement may   be assigned   by the Purchaser/the Tenant by way of absolute legal assignment to another person taking an assignment of the Purchaser's/the Tenant's interest in the Premises without the consent of the Client or the Firm being required and such assignment shall be effective upon written notice thereof being given to the Firm. No further assignment shall be permitted.]

(insert number of times)

(delete if under English law)

[7S. *The Purchaser/the Tenant shall   be entitled to assign or transfer his/their rights under this Agreement   to any other person acquiring the Purchaser's/the Tenant's interest in the whole of the Premises without the consent of the Firm subject to written notice of such assignation being given to the Firm in accordance with Clause 8 hereof. Nothing in this clause shall permit any party acquiring such right as assignee or transferee to enter into any further assignation or transfer to anyone acquiring subsequently an interest in the Premises from him.*]

8. Any notice to be given by the Firm hereunder shall be deemed to be duly given if it is delivered by hand at or sent by registered post or recorded delivery to the Purchaser/the Tenant at its registered office and any notice given by the Purchaser/the Tenant hereunder shall be deemed to be duly given if it is addressed to "The Senior Partner"/"The Managing Director" and delivered by hand at or sent by registered post or recorded delivery to the above-mentioned address of the Firm or to the principal business address of the Firm for the time being and, in the case of any such notices, the same shall if sent by registered post or recorded delivery be deemed to have been received forty eight hours after being posted.

(complete as appropriate)

9. No action or proceedings for any breach of this Agreement shall be commenced against the Firm after the expiry of   years from the date of practical completion of the Premises under the Building Contract.

(delete if under Scots law)

10. The construction validity and performance of this Agreement shall be governed by English law and the parties agree to submit to the non-exclusive jurisdiction of the English Courts.

[**AS WITNESS** the hands of the parties the day and year first before written.

(alternatives: delete as appropriate)

(for Agreement executed under hand and NOT as a Deed)

SIgned by or on behalf of the Firm ....

in the presence of: ....

Signed by or on behalf of the Purchaser/the Tenant ....

in the presence of: ....]

(this must only apply if the Appointment is executed as a Deed)

[**IN WITNESS WHEREOF** this Agreement was executed as a Deed and delivered the day and year first before written.

by the Firm

....

....

....

....

by the Purchaser/the Tenant

....

....

....

....]]

(delete if under English law)

10S. *This Agreement shall be construed and the rights of the parties and all matters arising hereunder shall be determined in all respects according to the Law of Scotland.*

***IN WITNESS WHEREOF** these presents are executed as follows:-*

SIGNED by the above named Firm at ....

on the .... day of .... Nineteen hundred and...............................

as follows:-

... (Firm's signature)

Signature .... Full Name ....

Address ....

.. Occupation ....

Signature .... Full Name ....

Address ....

.. Occupation ....

SIGNED by the above named Purchaser/Tenant at...

on the .... day of .... Nineteen hundred and...............................

as follows:-

For and on behalf of the Purchaser/the Tenant

.. Director/Authorised Signatory

.. Director/Authorised Signatory]

INDEX

Index

Index

By the same author:

Engineering Tenders, Sales & Contracts: Standard Forms and Procedures, E. & F.N. Spon and Sweet & Maxwell, 1982.

I.Mech.E/I.E.E. Conditions of Contract, Sweet & Maxwell, 1984.

Quick Reference to Forms